Sustainable Civil Infrastructures

About this Series

Sustainable Infrastructure impacts our well-being and day-to-day lives. The infrastructures we are building today will shape our lives tomorrow. The complex and diverse nature of the impacts due to weather extremes on transportation and civil infrastructures can be seen in our roadways, bridges, and buildings. Extreme summer temperatures, droughts, flash floods, and rising numbers of freeze-thaw cycles pose challenges for civil infrastructure and can endanger public safety. We constantly hear how civil infrastructures need constant attention, preservation, and upgrading. Such improvements and developments would obviously benefit from our desired book series that provide sustainable engineering materials and designs. The economic impact is huge and much research has been conducted worldwide. The future holds many opportunities, not only for researchers in a given country, but also for the worldwide field engineers who apply and implement these technologies. We believe that no approach can succeed if it does not unite the efforts of various engineering disciplines from all over the world under one umbrella to offer a beacon of modern solutions to the global infrastructure. Experts from the various engineering disciplines around the globe will participate in this series, including: Geotechnical, Geological, Geoscience, Petroleum, Structural, Transportation, Bridge, Infrastructure, Energy, Architectural, Chemical and Materials, and other related Engineering disciplines.

More information about this series at http://www.springer.com/series/15140

João Pombo · Guoqing Jing
Editors

Recent Developments in Railway Track and Transportation Engineering

Proceedings of the 1st GeoMEast International Congress and Exhibition, Egypt 2017 on Sustainable Civil Infrastructures

Springer

2017 GEOMEAST

SSIGE
SOIL-STRUCTURE INTERACTION GROUP IN EGYPT
المجموعة المصرية للهندسة الجيوتقنية والبنية التحتية

Editors
João Pombo
Heriot-Watt University
Edinburgh
UK

Guoqing Jing
Beijing Jiaotong University
Beijing
China

ISSN 2366-3405 ISSN 2366-3413 (electronic)
Sustainable Civil Infrastructures
ISBN 978-3-319-61626-1 ISBN 978-3-319-61627-8 (eBook)
DOI 10.1007/978-3-319-61627-8

Library of Congress Control Number: 2017946440

Printed on acid-free paper

This Springer imprint is published by Springer Nature
The registered company is Springer International Publishing AG
The registered company address is: Gewerbestrasse 11, 6330 Cham, Switzerland

Preface

Toward building sustainable and longer civil infrastructures, the engineering community around the globe continues undertaking research and development to improve existing design, modeling, and analytical capability. Such initiatives are also the core mission of the Soil-Structure Interaction Group in Egypt (SSIGE) to contribute to the ongoing research toward sustainable infrastructure. This conference series "GeoMEast International Congress and Exhibition" is one of these initiatives.

Ancient peoples built their structures to withstand the test of time. If we think in the same way, our current projects will be a heritage for future generations. In this context, an urgent need has quickly motivated the SSIGE and its friends around the globe to start a new congress series that can bring together researchers and practitioners to pursue "Sustainable Civil Infrastructures." The GeoMEast 2017 is a unique forum in the Middle East and Africa that transfers from the innovation in research into the practical wisdom to serve directly the practitioners of the industry.

More than eight hundred abstracts were received for the first edition of this conference series "GeoMEast 2017" in response to the Call for Papers. The abstracts were reviewed by the Organizing and Scientific Committees. All papers were reviewed following the same procedure and at the same technical standards of practice of the TRB, ASCE, ICE, ISSMGE, IGS, IAEG, DFI, ISAP, ISCP, ITA, ISHMII, PDCA, IUGS, ICC, and other professional organizations who have supported the technical program of the GeoMEast 2017. All papers received a minimum of two full reviews coordinated by various tracks chairs and supervised by the volumes editors through the Editorial Manager of the SUCI "Sustainable Civil Infrastructure" book series. As a result, 15 volumes have been formed of the final +320 accepted papers. The authors of the accepted papers have addressed all the comments of the reviewers to the satisfaction of the tracks chairs, the volumes editors, and the proceedings editor. It is hoped that readers of this proceedings of the GeoMEast 2017 will be stimulated and inspired by the wide range of papers written by a distinguished group of national and international authors.

Publication of this quality of technical papers would not have been possible without the dedication and professionalism of the anonymous papers reviewers. The names of these reviewers appear in the acknowledgment that follows. For any additional reviewers whose names were inadvertently missed, we offer our sincere apologies.

We are thankful to Dr. Hany Farouk Shehata, Dr. Nabil Khelifi, Dr. Khalid M. ElZahaby, Dr. Mohamed F. Shehata, and to all the distinguished volumes editors of the proceedings of the GeoMEast 2017. Appreciation is extended to the authors and track chairs for their significant contributions. Thanks are also extended to Springer for their coordination and enthusiastic support to this conference. The editors acknowledge the assistance of Ms. Janet Sterritt-Brunner, Mr. Arulmurugan Venkatasalam in the final production of the 15 edited volumes "Proceedings of GeoMEast 2017".

Contents

Experimental and Numerical Behavior of Railway Track Over Geogrid Reinforced Ballast Underlain by Soft Clay

Mohammed Y. Fattah(✉), Mahmood R. Mahmood, and Mohammed F. Aswad

Building and Construction Engineering Department, University of Technology, Baghdad, Iraq
myf_1968@yahoo.com
Mahoudal_qaisiy@yahoo.com

Abstract. In this paper, laboratory tests were conducted to investigate the effect of load amplitude, load frequency, on the behavior of reinforced and unreinforced ballast layer. A half full-scale railway was constructed for carrying out the tests, which consists of two rails 800 mm in length with three wooden sleepers (900 mm × 90 mm × 90 mm). The ballast was overlying 500 mm thickness clay. The tests were carried out with and without geogrid reinforcement; the tests were carried out in a well tied steel box of 1.5 m length × 1 m width × 1 m height. A series of laboratory tests were conducted to investigate the response of the ballast and the clay layers where the ballast was reinforced by a geogrid. Settlement in ballast and clay was measured in reinforced and unreinforced ballast cases.

In addition to the laboratory tests, the application of numerical analysis was made by using the finite element program PLAXIS 3D 2013 in order to verify the numerical model.

It was concluded that the amount of settlement increased with increasing the simulated train load amplitude, there is a sharp increase in settlement up to the cycle 500 and after that, there is a gradual increase to level out between, 2500 to 4500 cycles depending on the used frequency. There is a little increase in the induced settlement when the load amplitude increased from 0.5 to 1 ton but it is higher when the load amplitude increased to 2 ton, the increasing amount in settlement depends on the geogrid existence and the other studied parameters. Both experimental and numerical results have the same behavior. For unreinforced case, it was observed that the experimental results at the beginning show higher readings than the numerical ones, but after a number of cycles ranging from 750 to 2500 cycles the numerical results show higher readings.

It was also concluded that increasing the ballast layer thickness from 20 cm to 30 cm leads to decrease the settlement by about 50%. This ascertains the efficiency of ballast in spreading the waves induced by the track. The effect of load frequency on the settlement ratio is almost constant after 500 cycles.

Keywords: Railway track · Geogrid reinforcement · Finite elements · Ballast · Soft clay

J. Pombo and G. Jing (eds.), *Recent Developments in Railway Track and Transportation Engineering*, Sustainable Civil Infrastructures, DOI 10.1007/978-3-319-61627-8_1

1 Introduction

Railway ballast forms a major component of a conventional rail track and is used to distribute the load to the subgrade, providing a smooth running surface for the train. It plays a significant role in providing support for the rail track base and distributing the load to the underneath weaker layer. Ballast also helps with drainage, which is an important factor for any type of transportation structure, including railroads. Over time ballast deforms and degrades progressively under dynamic loading and loses its strength.

Despite the problems associated with ballast, ballast is still a preferable choice for substructure material over other alternatives such as concrete slabs or asphalt. This is because ballast provides less stiff support which is an important factor in case of differential settlement or subgrade failure, ballast is more economical, produces less noise (Profillidis 2000), contributes in damping impact wheel load and railway track vibration, distributes transmitted stresses evenly, reduces the settlements and provides a good base layer under low confining pressures. However, there are many cases where using ballast individually may not be sufficient because of subgrade layer weakness.

To stabilize the tracks encountering weak foundation soils and ballast degradation under high-speed loadings, the ballast is reinforced with geosynthetics. Some of the functions of geosynthetics include separating the ballast layer from the subgrade to prevent ballast penetration, reinforcing the ballast layer to reduce ballast settlement and attrition, filtration to prevent subgrade pumping and drainage to prevent excessive wetting of the subgrade. These techniques can reduce the depth of the required granular layer and also reduce the frequency of the required maintenance.

Since past few decades, geosynthetics have been increasingly used in the construction of various geotechnical structures, like earth retention, slope stability, roadway and railway construction, due to its ease of use and cost efficiency, geosynthetics has been developed in many forms and materials. These include geogrids, geomembranes, geotextiles, geocomposites and geocells (Koerner and Koerner 2005).

Indraratna et al. (2006) carried out a series of tests to measure the parameters (settlement, vertical strain, and lateral strain) of reinforced (geogrid-geotextile) recycled ballast in wet and dry conditions. Using of geogrid-geotextile showed an increase in the bearing capacity and resilient modulus of recycled ballast. The results also show a decrease in degradation and lateral movement of ballast. A numerical analysis (finite element) was conducted to determine the optimum depth. It was found that optimum depth for geosynthetics is in the range of 150 to 200 mm beneath the sleepers.

Heidari and El Naggar (2010) investigated the effect of the soil reinforcement on the performance characteristics for different configurations of shock absorbing foundations. The results showed that soil reinforcement can be used to increase the stiffness of the supporting medium. This increase can be designed to achieve a superior dynamic performance for shock-producing equipment when the mounting system alone cannot achieve a satisfactory design. It was shown that for small hammers, the reinforced soil foundation can reduce the foundation response amplitude by up to 80%. For large hammers and presses, the reinforced soil foundation can be designed to reduce the foundation response by up to 60% of the case of no soil reinforcement.

Kwan (2006) carried out a Composite Element Test (CET). The aim of the test was to examine various variables (subgrade stiffness, geogrid properties, and geogrid location in the reinforced ballast), he found that the aperture size of the geogrid has a direct effect on the behavior of a reinforced composite settlement characteristic. The results showed that, the geogrid aperture size ratio of 1.6 to particle diameter gave the best performance with aggregate diameter of 40 mm.

Sowmiya et al. (2011), studied geosynthetic reinforced railway tracks model using ABAQUS 6.9 with different sub ballast thicknesses and compared with unreinforced section. The results showed that the reinforcement can be used to improve the performance of railway tracks on clayey subgrade. The study showed that the reinforcement between sub ballast and subgrade, between ballast and Sub ballast and the reinforcement at both the interfaces reduce induced vertical stresses and displacements significantly. The conclusion of the study was that to reduce the maintenance cost and to reduce the shear failure, the reinforcement between sub ballast and subgrade, between ballast and sub ballast and the reinforcement at both the interfaces are the best options.

Nguyen et al. (2011), worked to develop the most adequate and efficient models for calculation of dynamic traffic load effects on railways track infrastructure, and then evaluate the dynamic effect on the ballast track settlement, using a ballast track settlement prediction model, which consists of the vehicle/track dynamic model previously selected and a track settlement law. The calculations were based on dynamic finite element models with direct time integration, contact between wheel and rail and interaction with railway cars. An initial irregularity profile is used in the prediction model. The track settlement law was considered to be a function of number of loading cycles and the magnitude of the loading, which represents the long-term behavior of ballast settlement. The results obtained included the track irregularity growth and the contact force in the final interaction of numerical simulation. They included that an increase of train speed will produce higher contact forces between the wheel and the rail, and will produce larger deflections in the ballast and a larger settlement will be obtained.

Wayne et al. (2013) conducted a controlled field study in Weirton, West Virginia, USA; to evaluate performance of a geogrid stabilized unpaved aggregate base overlying relatively weak and non-uniform subgrade soils. The results showed that the horizontal pressures within the subgrade created by both the static and live loading conditions were significantly reduced by using the geogrid. Also results confirmed that the geogrid improved aggregate confinement and interaction, leading to enhanced structural performance of the unpaved aggregate base.

Leshchinsky and Ling (2013), studied a numerical modeling using finite element analysis according to experimental results of large-scale laboratory tests of geocell reinforced ballast embankments for confinement. The study was to explore (the geocell confinement effects on ballasted embankments resting on soft subgrade, different ballast stiffnesess, and varying stiffness of reinforcement). From the results, it was found that geocell confinement has a significant benefit on a wide range of subgrade stiffness. The effect of the confined ballast is distributing stresses more uniformly to the subgrade, which can provide higher bearing capacities and less settlement.

Railroad ballast layer consists of discrete aggregate particles and Discrete Element Method (DEM) is one of the most suitable ways to simulate the deformation behavior of particulate nature of ballast materials. An aggregate imaging based DEM simulation platform developed at the University of Illinois at Urbana–Champaign (UIUC) can simulate railroad ballast behavior through the use of polyhedron shaped discrete elements. These ballast elements were created by Tutumluer et al. (2007, 2014) with realistic size and shape properties from image analyses of actual particles using an Aggregate Image Analyzer. The UIUC railroad ballast DEM model was recently put to test for predicting settlement behavior of full-scale test sections under repeated heavy axle train loading. Field settlement data were collected from the Facility for Accelerated Service Testing (FAST) for Heavy Axle Load (HAL) applications at Transportation Technology Center (TTC) in Pueblo, Colorado, to validate the DEM model. The ballast settlement predictions due to the repeated train loading indicate that the DEM model could predict magnitudes of the field ballast settlements from both early loading cycles and over 90 Million Gross Tons (MGTs) performance trends reasonably accurately. The settlement predictions were sensitive to aggregate shape, gradation and initial compaction condition (density) of the constructed ballast layer.

The main objective of the present research is to investigate the advantages of utilizing geogrid as reinforcement to minimize the amount of degradation of ballast under railway track base resting on clayey soil, to determine the transmitted stresses and the developed excess pore water pressure in the clay layer due to dynamic load and to investigate the amount of resilient settlement happened in the track model.

In addition to the laboratory tests, the application of numerical analysis was made using the finite element program PLAXIS 3D 2013. A laboratory model simulations was carried out in order to verify the numerical model and for the comparison between theoretical and experimental results.

1.1 Model Preparation and Experimental Works

Laboratory tests were conducted to investigate the response of the ballast and the clay layers where the ballast was reinforced by a geogrid. Settlement in ballast and clay, soil pressure and pore water pressure induced in the clay layer were measured in reinforced and unreinforced ballast cases.

Special testing apparatus and other accessories were designed and manufactured to simulate the condition as close as possible to those occurring in the field. The apparatus has the capability of applying different dynamic loads under different frequencies. The general view of the apparatus is shown in Figs. 1 and 2.

The apparatus consists of the following items: 1. Loading steel frame. 2. Hydraulic loading system. 3. Load spreader beam. 4. Data acquisition. 5. Shaft encoder. 6. Steel container.

Steel loading frame: The steel loading frame shown in Fig. 2 was designed to restrain the vertical hydraulic load jack which is applying the dynamic loading to mimics the passing of train. The hydraulic jack was bolted to the top of two beams of the loading frame by plates welded under them.

Fig. 1. General view of the loading system.

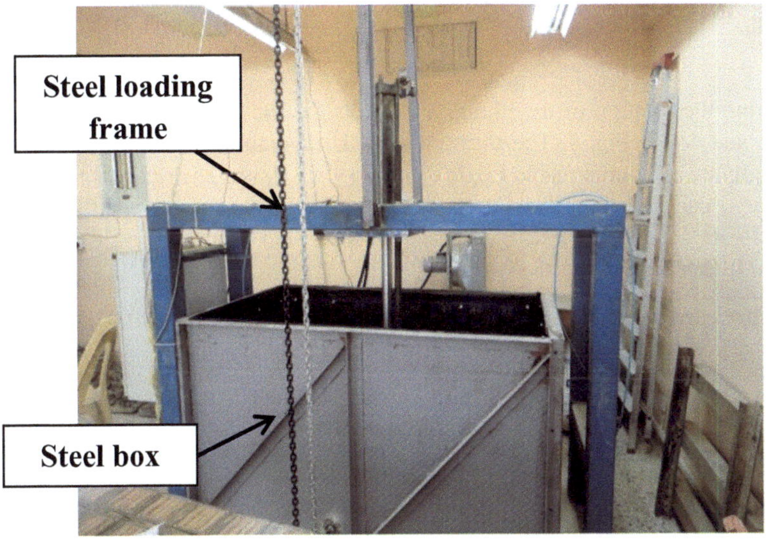

Fig. 2. Steel frame of the loading system.

Hydraulic loading system: The system contains a hydraulic steel tank with a capacity of 70 L. The tank includes a gear type hydraulic pump. The axis of the pump is connected by a coupling with a three phase electrical rotary motor of 3 hp capacity and 1450 rpm rotation speed. The pump and the motor are fixed in housing on the upper surface of the tank Fig. 3.

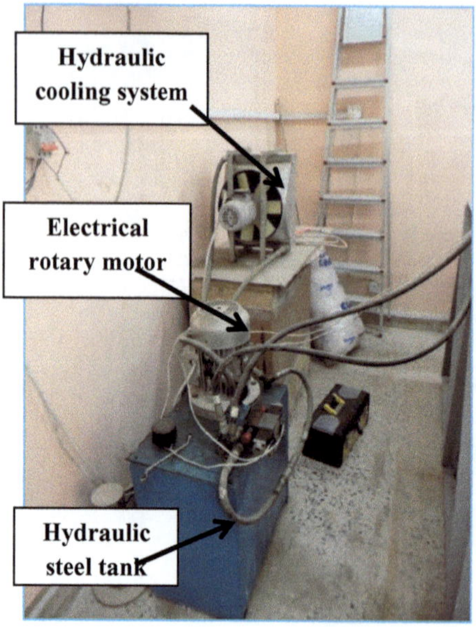

Fig. 3. Hydraulic loading system.

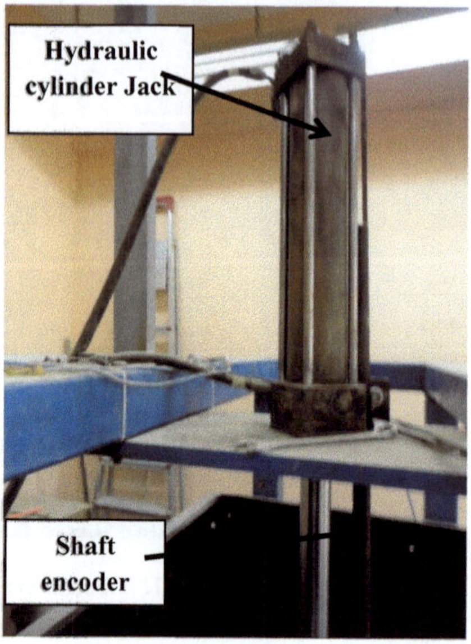

Fig. 4. Hydraulic cylinder jack.

During the rotation of the pump, it controls the moving of the hydraulic cylinder jack up and down, Fig. 4. The movement of the hydraulic cylinder jack is controlled electrically by a Programmable Logic Control (PLC) through which, the movement (up and down) can be controlled by choosing the hearts that are needed in the control through data acquisition system. The data acquisition system also displays the load magnitude that applied on the rail.

Load spreader beam: An 800 mm × 50 mm × 50 mm solid steel beam was used to apply the load on the track panel as shown in Fig. 5.

The steel container: The tests were carried out in a steel container with a plan dimension of 1.5 m length × 1 m width × 1 m height. Each part of the container was made of a steel plate 5 mm thick.

Data acquisition system: Data acquisition system is used to measure and sense the occurring displacement during the tests, which enable the tester to obtain a huge data of readings in a very short time, moreover it is used to choose the specified frequency used in the test.

The data acquisition system consists of Programmable Logic Controller (PLC) which can be defined as a digital computer used for electro-mechanical automation processes, and it is a high technology processing unit. This type of systems analyzes the data digitally.

Fig. 5. Load spreader beam.

Shaft encoder: A shaft encoder is an electro-mechanical device used to convert the motion of the shaft to a digital code. The output of incremental encoder supply information of the motion of the shaft which is processed into information such as, displacement, revolution per minute (rpm), speed, and position.

Used Materials

A brown clayey soil was used. Standard tests were performed to determine the physical properties of the clay. Details are given in Table 1. According to the Unified Soil Classification System U.S.C.S, the soil is classified as (CL).

The ballast was produced as a result of crushing big stones; the ballast is of white color with angular shapes. The effective size, uniformity coefficient and coefficient of gradation are listed in Table 2. The ballast is of uniform size with poorly graded gradation (GP) according to the Unified Soil Classification System.

The geogrid used in all tests was manufactured by Tensar type SS2, Fig. 6. The selected geotextile must meet the following four durability criteria.

(a) It must be tough to withstand the stresses during the installation process. Properties concerned are: Tensile strength, burst strength, grab strength, tear strength, resistance to ultraviolet (UV) light degradation for two weeks exposure with negligible strength loss.

(b) It must be strong enough to withstand static and dynamic loads, high pore pressures, and severe abrasive action to which it is subjected during its life-span. The properties concerned are: Puncture resistance, abrasion resistance, longation at failure.

Table 1. Physical properties of clay used

Property	Value	Specification
Liquid limit (LL)	46	ASTM D 4318
Plastic limit (PL)	21	ASTM D 4318
Plasticity index	25	ASTM D 4318
Specific gravity (Gs)	2.65	ASTM D 854
Gravel%, >4.75 mm	0	ASTM D 422
Sand%, 0.075–4.75 mm	4	ASTM D 422
Silt%, 0.005–0.075 mm	35	ASTM D 422
Clay%, <0.005 mm	61	ASTM D 422
D_{85} mm	0.018	ASTM D 422
D_{60} mm	0.0036	ASTM D 422
D_{30} mm	–	ASTM D 422
D_{15} mm	–	ASTM D 422
D_{10} mm	–	ASTM D 422
Activity	0.41	ASTM D 4318
Coefficient of uniformity, Cu		ASTM D 422
Coefficient of gradation, C_c		ASTM D 422

Table 2. Ballast particle size characteristics

Parameter	Value
D_{60}	21.59 mm
D_{30}	20.61 mm
D_{10}	18.35 mm
Coefficient of uniformity, C_u	1.18
Coefficient of gradation, C_c	1.07
$\gamma_{dry\ min}$	15.21 kN/m^3
$\gamma_{dry\ max}$	19.25 kN/m^3
γ_n	17.83 kN/m^3
Relative density	70%

(c) It must be resistant to excessive clogging or blinding, allowing water to pass freely across and within the plane of the geotextile. At the same time, it must be capable of filtering out and retaining fines in the subgrade. The properties concerned are: Cross-plane permeability or permittivity, in-plane permeability or transmissivity, apparent opening size (AOS).

Raymond and Bathurst (1990) have shown that abrasion resistance is a function of the opening size. Geotextiles having low opening size are more abrasion resistant due to the fact that smaller number of particles with maximum size would be able to penetrate the geotextile. The degree of needle-punching that a non-woven needle-punched geotextile receives during manufacture determines the amount of interlock between fibers and hence, influences the geotextile abrasion resistance.

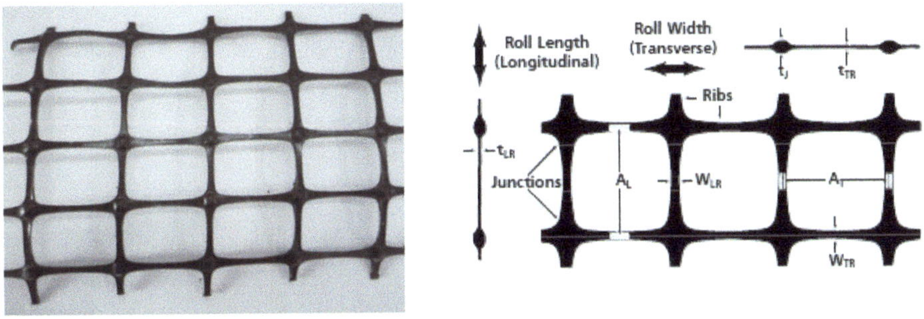

Fig. 6. Geogrid reinforcement used in tests.

Fig. 7. Track panel.

A track panel that consists of two rails 80 cm in length and three wooden sleepers (900 mm × 100 mm × 100 mm) was used in the tests as shown in Fig. 7. Figure 8 shows the dimensions of the rail used in the tests in contrast with the real rail dimensions. The spacing between the rails and the sleepers are 650 mm and 300 mm center to center, respectively.

1.2 Preparation of Test Model

A half full-scale railway was constructed for carrying out the tests. Two rails 80 cm in length with three wooden sleepers (900 mm × 100 mm × 100 mm) were used to construct the track panel, Fig. 9. The ballast was overlying 50 cm thickness soft clay. The tests were carried out with and without geogrid reinforcement.

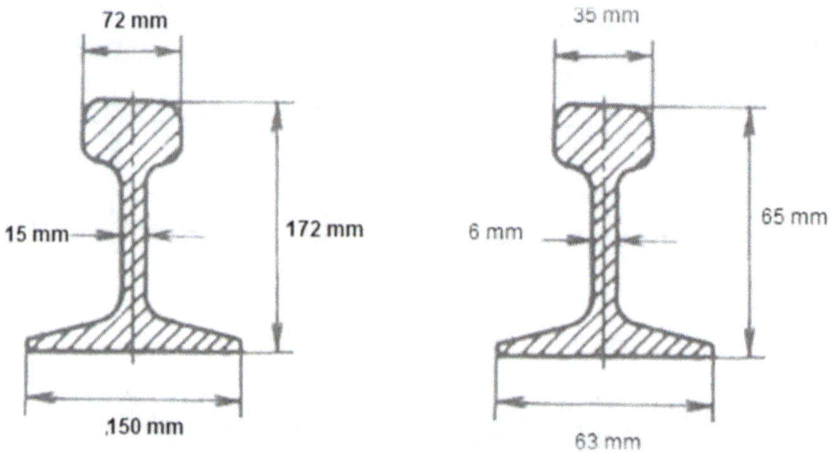

a. Real rail dimension. b. Model rail used in experiment.

Fig. 8. Dimensions of the rail (Li et al. 2016).

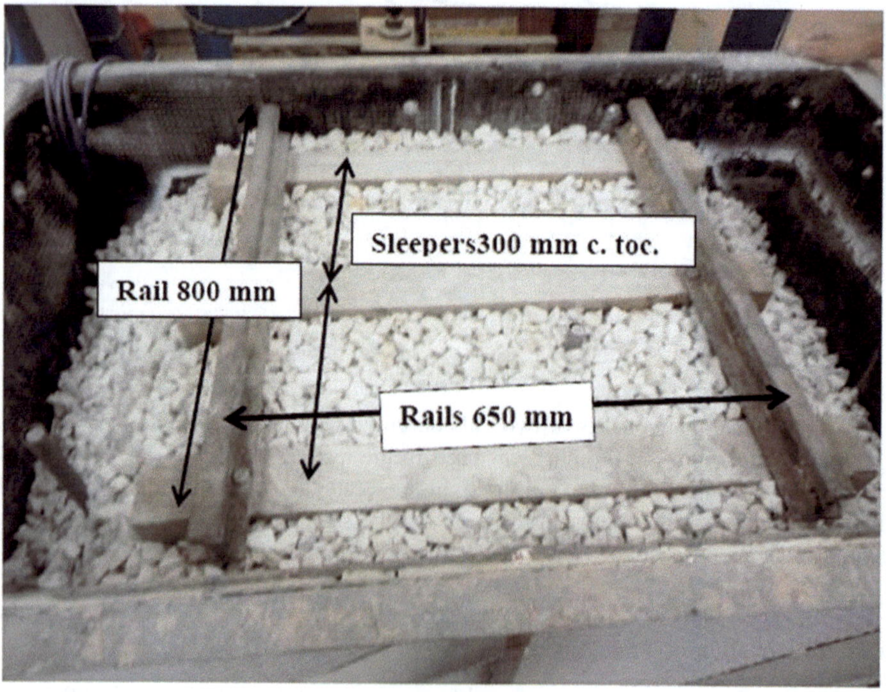

Fig. 9. Track panel dimensions.

Prior to the stage of preparation of the bed of soil (clay), control tests were carried out to determine the variation of the undrained shear strength with time at different water contents (25%, 28%, 32%, 35% and 40%) to specify the time required for the remolded soil to regain its strength after a rest period following the mixing process. It was found that the time required for the remolded soil to regain its strength is about 3 days.

Ballast Layer
The construction of the ballast layer starts after three days from the preparation of the soil bed. The ballast is placed carefully on the surface of the soil bed (clay) in layers; each layer is not more than 100 mm thick. Three values of ballast thicknesses of; 200, 300 and 400 mm were used in the tests, each side of the ballast was sloped down on about 2:1 slope. A predetermined volume of ballast was prepared which is sufficient to create a uniform layer. Each layer was compacted gently by a tamping rod to attain a placement dry unit weight of about 17.83 kN/m^3. This placement unit weight corresponds to a relative density of about 70%.

1.3 Test Procedure

The test was carried out in a well tied steel box of 1500 mm length × 1000 mm width × 1000 mm height, the box was padded with two layers, the first one consists of compressed styropor sheets 5 mm thick and the other one is a rubber 4 mm thick to prevent reflection of waves during the test.

The box was filled with relatively soft clay which was placed in 100 mm layers to ensure the consistency and was compacted by plywood to a depth of 500 mm. After the placement of each layer, it was pressed gently with a wooden tamper in order to remove entrapped air.

The clay material used in the tests has a wet unit weight of 19.5 kN/m^3, moisture content of about 30% and undrained shear strength of about 25 kN/m^2 when it was finally placed in the box in its soft state and of a wet unit weight 21.6 kN/m^3, moisture content of about 21%.

In the laboratory, ballast was hoisted and placed over the clay layer in the box by using 2 ton manual chain hoists, where the ballast was placed in plastic containers and then the ballast is thrown carefully to allow the ballast to fall into the box in a controlled manner. A predetermined weight of ballast was prepared in accordance with the expected volume and the density used in the tests. The ballast was placed and compacted in 100 mm layers by using a tamping rod to attain a dry unit weight of 17.83 kN/m^3 corresponding to a relative density of about 70% to depths of 200 mm or 300 mm or 400 mm as the test required. The side slope of the ballast layer was controlled by maintaining alignment with markings on the wall of the box.

During preparation of the ballast layer, the geogrid was laid in a predetermined position as specified in the reinforced ballast tests.

Then the track panel was placed into its particular position via using the manual chain hoists and a good seating on the surface of the ballast was achieved by tamping it

Fig. 10. Lowering the load spreader beam to touch the track.

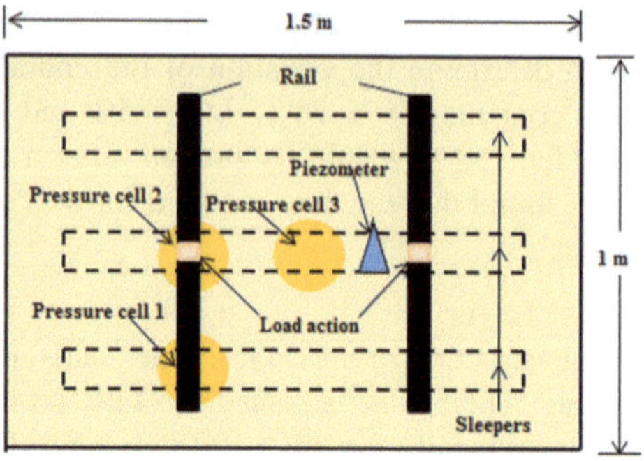

Fig. 11. Pressure cell distribution layout.

carefully. Special care was given to the leveling of the track panel and sleeper at the position where the rail must be placed. Additional amount of ballast was then added between the sleepers and at ends of the sleepers to achieve restraint.

After preparing the model for test, it was carried and centered in the load setup and then the load spreader beam was lowered carefully until it touched the rails. After that, the load and frequency were adjusted by the LCD touch-screen as required and then the test was started, Fig. 10.

Vertical settlements were measured by a displacement transducer which was built within the hydraulic jack body to measure the movement under the shaft; this settlement represents the average settlement for the entire track panel. Settlements on the ends of the outer sleeper were measured by a linear vertical displacement transformer (LVDT).

Stress transmitted and excess pore water pressure developing from the applied load were recorded by three pressure cells installed on the clay surface and a piezometer installed at 50 mm depth in the clay layer, they were placed at specific locations to measure the stresses transmitted by the ballast to the clay layer. The layout of the cell pressure distribution is shown in Fig. 11, two of the cells were placed under one of the ends and the center of the central sleeper and the third one is placed at the end of one of the outer sleepers.

The traffic loading simulation on the sleepers was executed by applying rectified sine wave loading. This type of loading was suggested by Awoleye (1993). It simulates a running of train over three sleepers in which 50% of the wheel load is transmitted to the middle sleeper and 25% of the wheel load on both outer sleepers.

The effect of applying geogrid reinforcement was examined by the comparison of settlement magnitude, the pressure and pore water pressure developed in the clay layer performed in various tests with and without geogrid in the ballast layer.

2 Finite Element Simulation

An application of numerical analysis was made using the finite element program PLAXIS 3D 2013 in order to verify the numerical model and for the comparison between theoretical and experimental results. Table 3 lists the material properties used in the analysis.

Table 3. Material properties used in the numerical analysis

Parameter	Clay	Ballast	Sleeper (timber)	Rail (steel)
Material model	Hardening soil	Mohr-Coulomb	Linear elastic	
Drainage type	Undrained	Drained		
Unit weight kN/m^3	19.5	17.83	9	78.5
Modulus of elasticity kN/m^2		110×10^3	7.2×10^6	205×10^6
Cohesion S_{u}, c	25	1	–	–
Friction angle ϕ	0	45°	–	–
Dilatancy anglc, Ψ	0	10	–	–
Poison's ratio, ν	0.449	0.35	0.3	0.28
$e_{initial}$	0.61	0.7	0.5	0.5
Compression index, C_c	0.18	–	–	–
Swelling index C_s	0.1	–	–	–
Gcogrid normal elastic stiffness EA kN/m		600		

The ballast is assumed to follow a Mohr-Columb failure criterion. The rail and sleepers are modeled as linear elastic, non-yielding behavior material and high magnitude of stiffness of these materials is assumed in comparison to those of the ballast. Clayey soils are modeled as a hardening undrained soil without considering any time-dependent behavior, such as consolidation. The geogrid was modeled as an elastic material.

2.1 Calculation Phases and Boundary Condition

The calculation consists of three phases except the initial phase for generating the initial stresses with active groundwater table. The process of setting the ballast was chosen in phase one. Phase two was to simulate the elements of the railway track (sleepers and rail). The dynamic load was selected in phase three to consider settlement and stresses in the soil.

To simulate the plane strain boundary condition (y direction) as in the test model, two plates were constructed in the xz plane at ballast cross section at the minimum and maximum y direction to prevent the ballast movement in this direction. Interface surface between the plate and the ballast was added to allow ballast movement in x and z direction.

Table 4. Tests identification for **models on soft clay**

Test no.	Test name identification	Ballast thickness, cm	Load amplitude, ton	Load frequency, Hz	No. of layers	Layer position (h/T)
10	T30 A2 f2 NL1-0.25	30	2	2	1	0.25
26	T30 A0.5 f1 NL1-0.25	30	0.5	1	1	0.25
37	T20 A2 f1 NL0	20	2	1	0	–
38	T20 A2 f1 NL1-0.25	20	2	1	1	0.25
13	T30 A1 f1 NL0	30	1	1	0	–
9	T30 A2 f2 NL0	30	2	2	0	–
14	T30 A1 f1 NL1-0.25	30	1	1	1	0.25

where:
T: ballast layer thickness for soft clay tests, cm,
DT: ballast layer thickness with sand drains tests, cm,
A: load amplitude, ton,
F: load frequency, Hz,
NL: number of geogrid layers and geogrid layer position (h/T) in ballast layer, and
h: position of geogrid layer in ballast layer from the upper surface of the clay layer

Table 4 lists the experimental tests that are carried out. To facilitate the discussion of the results, tests are given symbolic names; these names will be adopted under the variables in the experiments as shown in the table.

2.2 Results of Comparison Models and Discussion

The comparison will be carried out for the model tests, T20 A2 f1 NL0, T20 A2 f1 NL1-0.25, T30 A1 f1 NL0, T30 A2 f1 NL1-0.25, T30 A2 f2 NL0, and T30 A2 f2 NL1-0.25. Figures 12, 13, 14, 15, 16 and 17 present the settlement versus number of cycles relationship for experimental and numerical results.

From the figures, it can be observed that, both experimental and numerical results have the same behavior. For unreinforced case, it was observed that the experimental results at the beginning show higher readings than the numerical ones, but after a number of cycles ranging from 750 to 2500 cycles, the numerical results show higher readings. At the same time, it can be observed that, for reinforced case, the experimental results are always higher than the numerical.

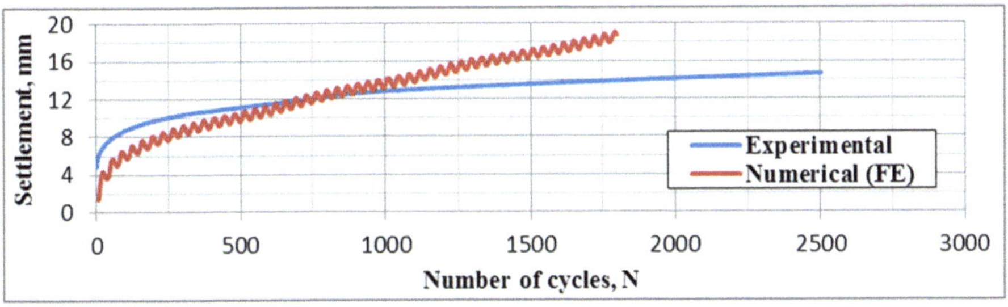

Fig. 12. Comparison between the measured and predicted settlement versus number of cycles for test T20 A2 f1 NL0.

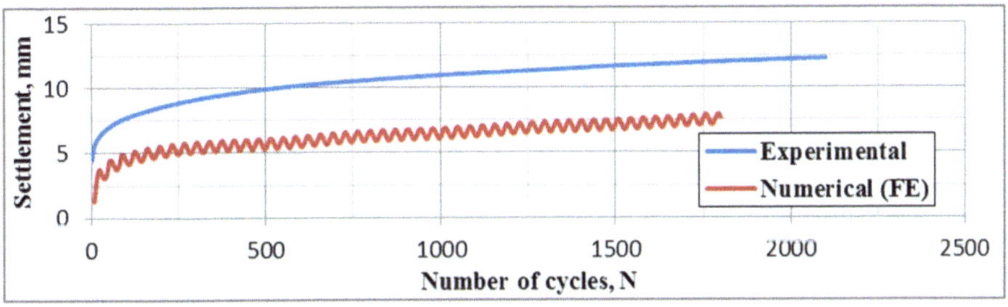

Fig. 13. Comparison between the measured and predicted settlement versus number of cycles for test T20 A2 f1 NL1-0.25.

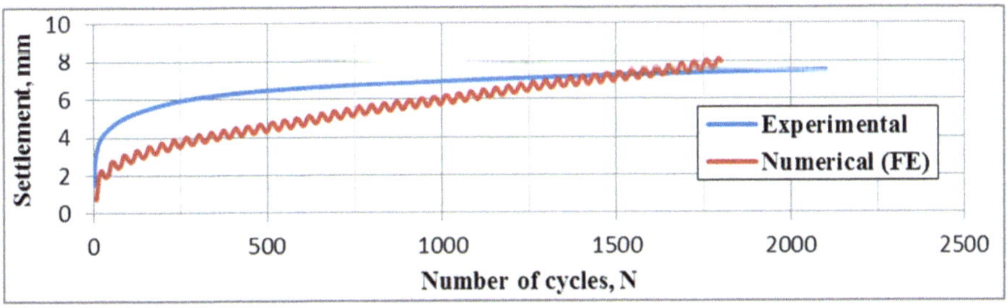

Fig. 14. Comparison between the measured and predicted settlement versus number of cycles for test T30 A1 f1 NL0.

Figures 18, 19, 20, 21 and 22 show the developed pressure and pore water pressure with number of cycles relationship for experimental and numerical results. From the figures, it can be observed that the experimental tests almost have higher readings and a wider range of readings.

Inclusion of reinforcement will redistribute the applied load to a wider area, thus minimizing stress concentration and achieving a more uniform stress distribution. Placement of a geogrid layer or layers in or at the bottom of the ballast course allows for shear interaction to develop between the ballast and the geogrid, as the base

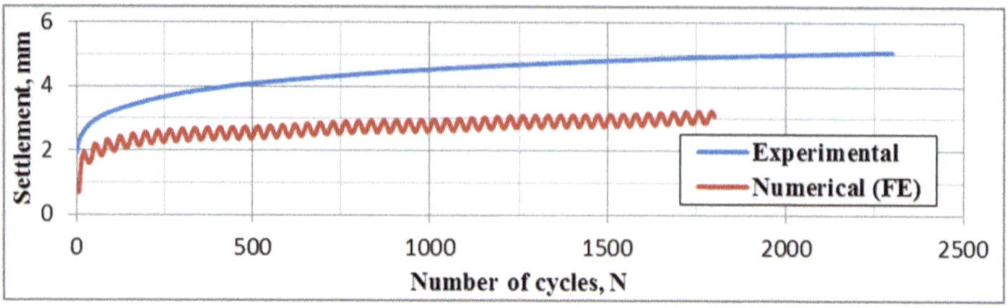

Fig. 15. Comparison between the measured and predicted settlement versus number of cycles for test T30 A1 f1 NL1-0.25.

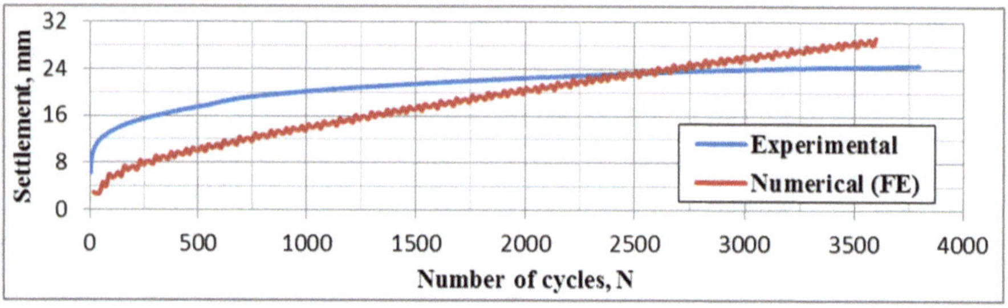

Fig. 16. Comparison between the measured and predicted settlement versus number of cycles for test T30 A2 f2 NL0.

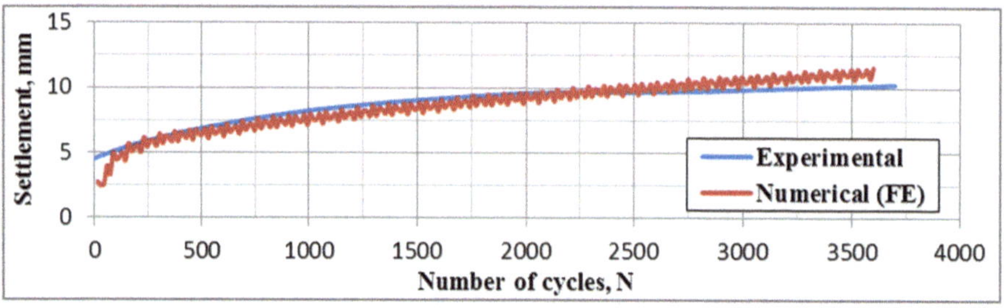

Fig. 17. Comparison between the measured and predicted settlement versus number of cycles for test T30 A2 f2 NL1-0.25.

attempts to spread laterally. Shear load is transmitted from the ballast to the geogrid and places the geogrid in tension. The relatively high stiffness of the geogrid acts to retard the development of lateral tensile strain in the ballast adjacent to the geogrid. Lower lateral strain in the ballast results in less vertical deformation of the track way surface. Hence, the first mechanism of reinforcement corresponds to direct prevention of lateral spreading of the ballast.

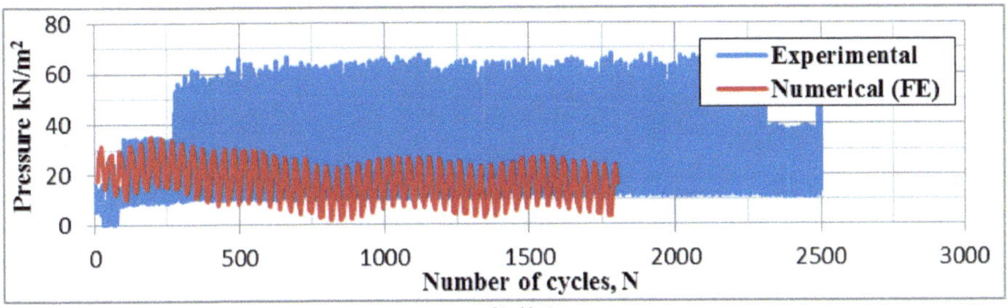

a- Cell 1

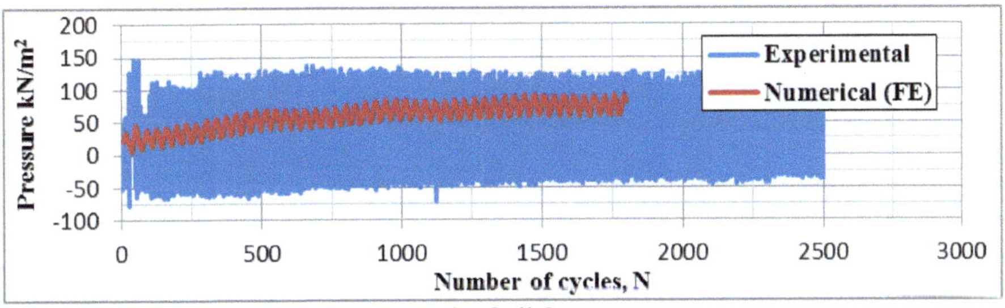

b- Cell 2

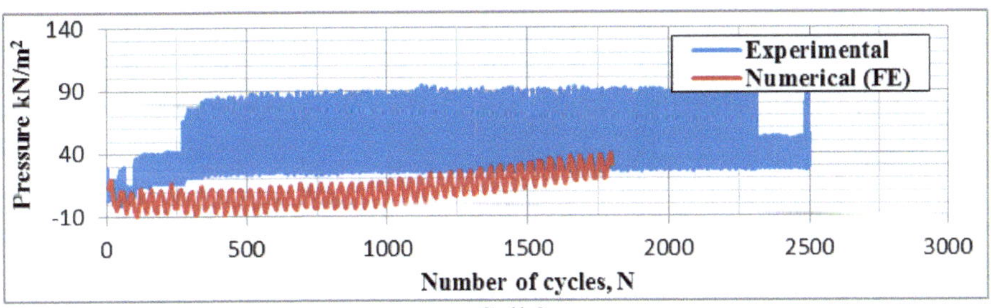

c- Cell 3

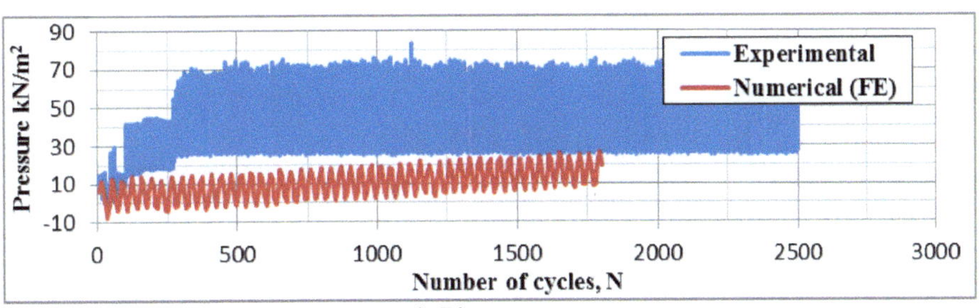

d- Piezometer

Fig. 18. Comparison between the measured and predicted pressure with number of cycles for test T20 A2 f1 NL0.

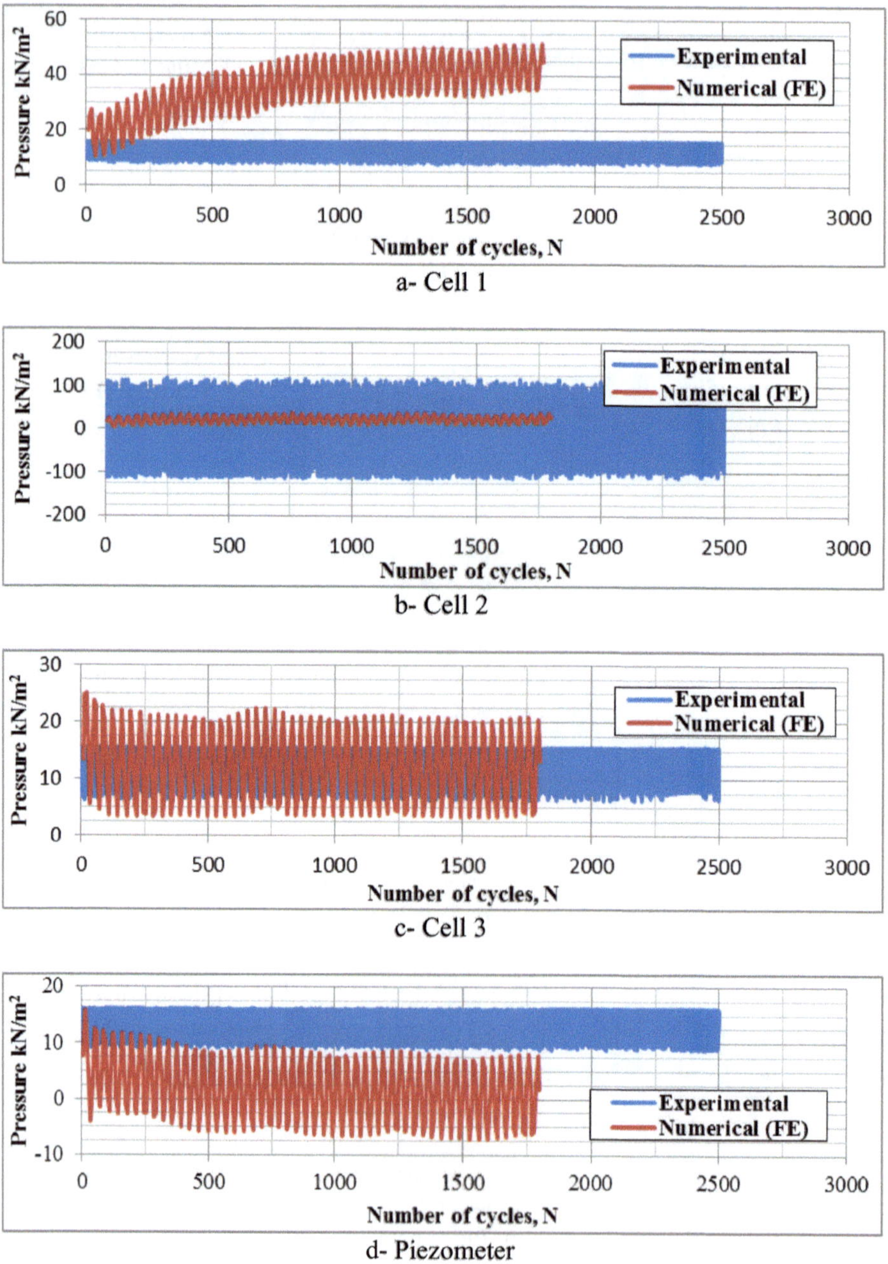

Fig. 19. Comparison between the measured and predicted pressure with number of cycles for test T20 A2 f1 NL1-0.25

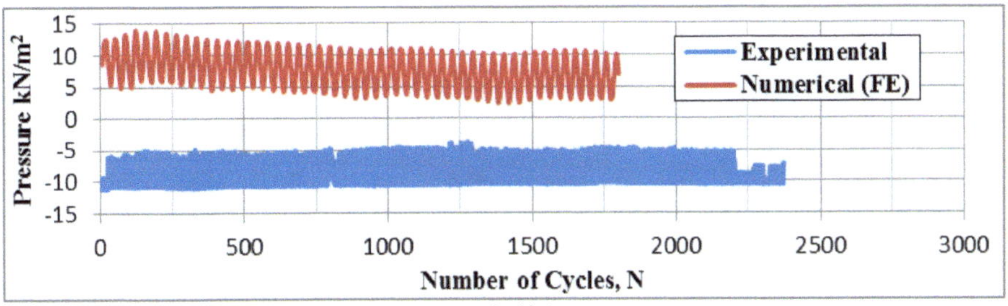

a- Cell 1

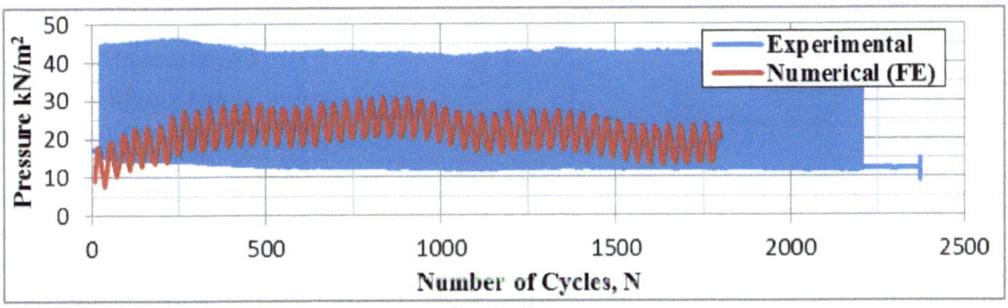

b- Cell 2

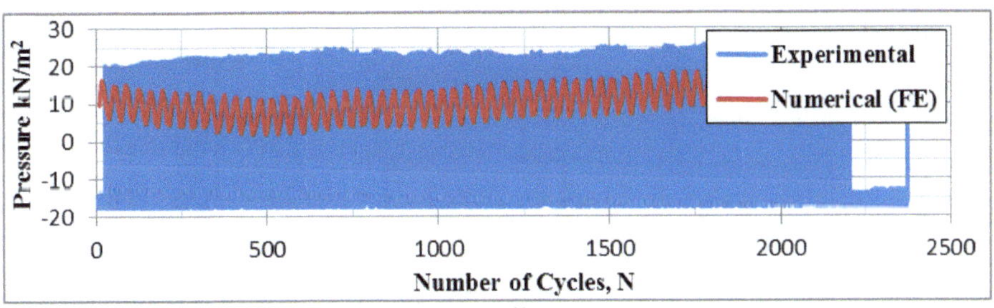

c- Cell 3

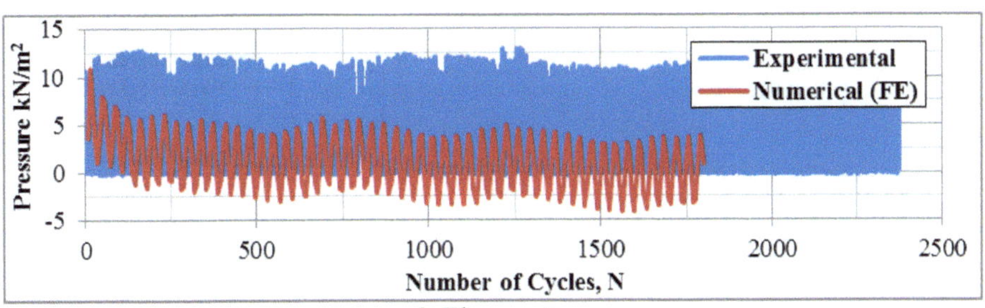

d- Piezometer

Fig. 20. Comparison between the measured and predicted pressure with number of cycles for test T30 A1 f1 NL0

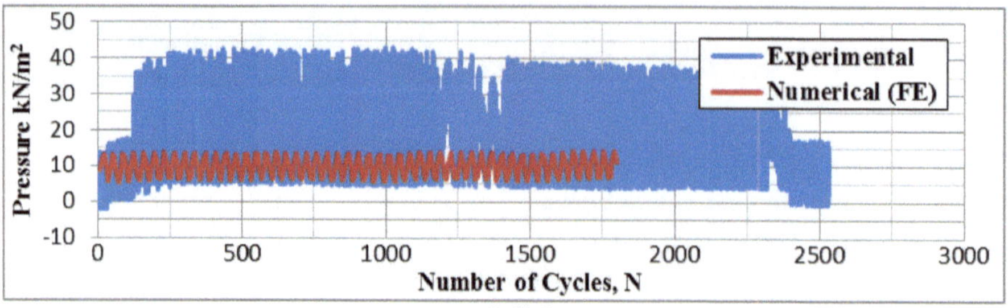

a- Cell 1

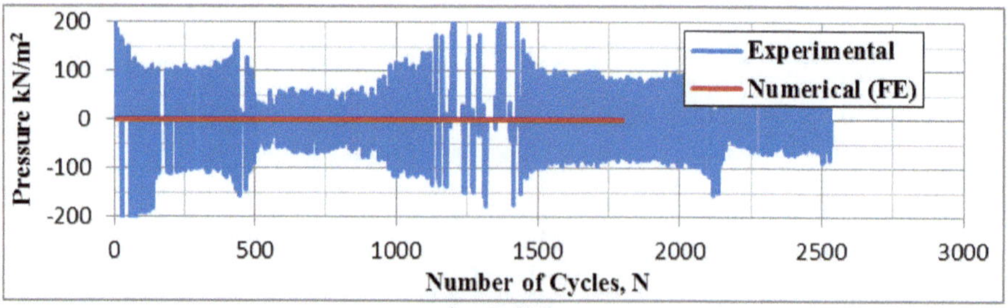

b- Cell 2

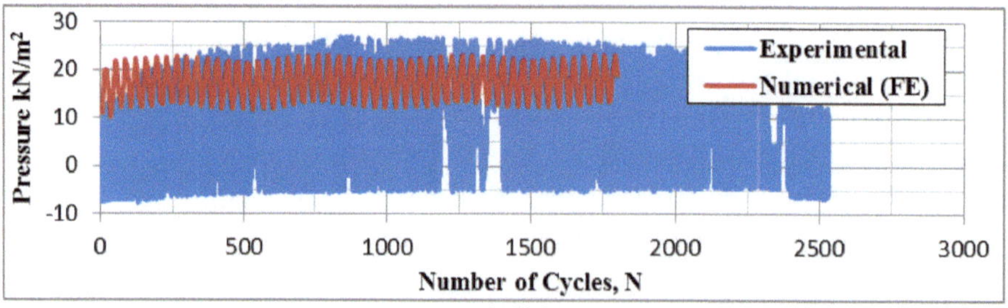

c- Cell 3

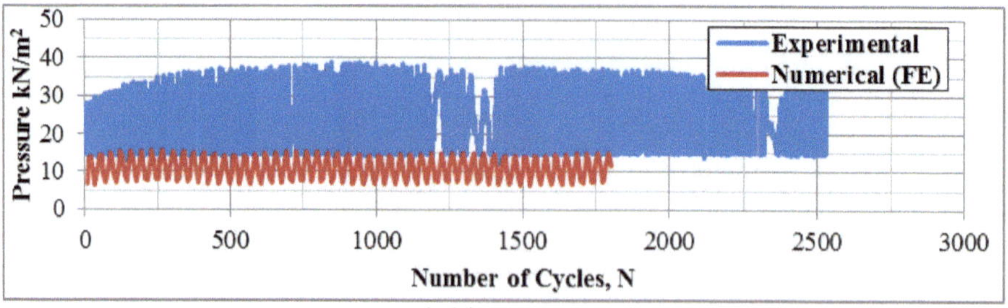

d- Piezometer

Fig. 21. Comparison between the measured and predicted pressure with number of cycles for test T30 A1 f1 NL1-0.25

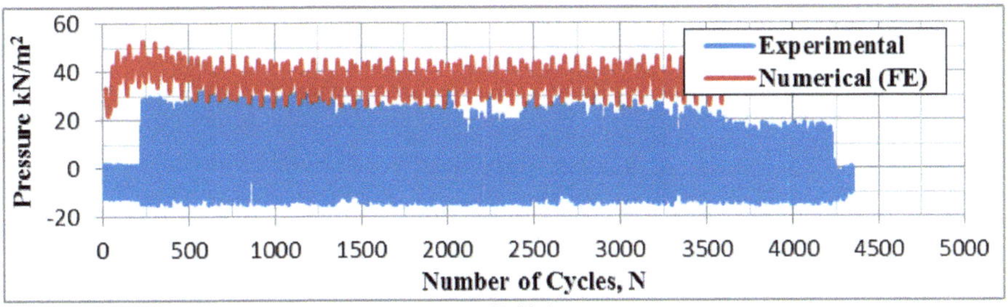

a- Cell 1

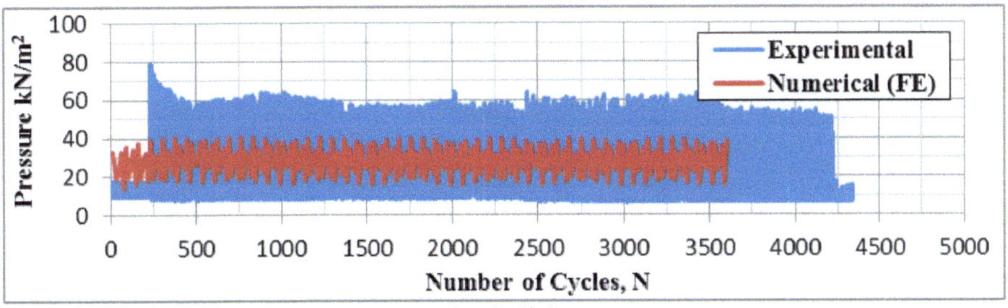

b- Cell 2

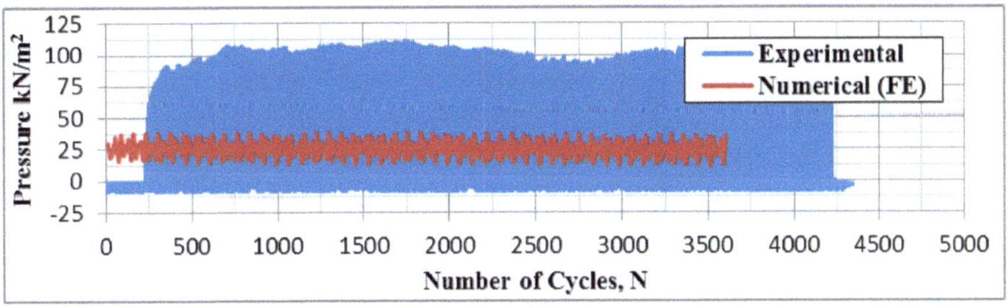

c- Cell 3

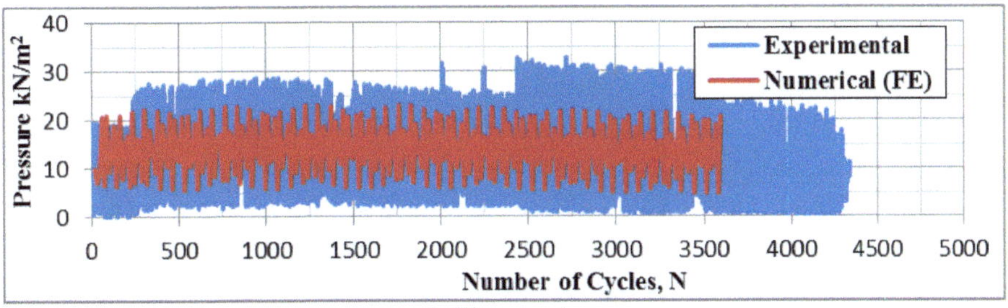

d- Piezometer

Fig. 22. Comparison between the measured and predicted pressure with number of cycles for test T30 A2 f2 NL1-0.25.

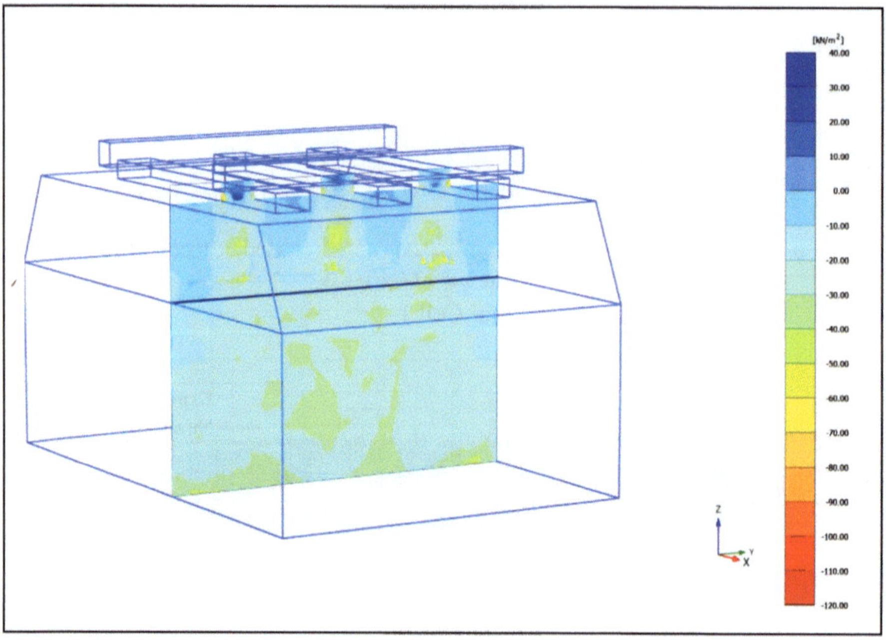

Fig. 23. Total stresses (kPa) in z direction vertical section at the end of phase three for the test model T30 A2 f1 NL1-0.25.

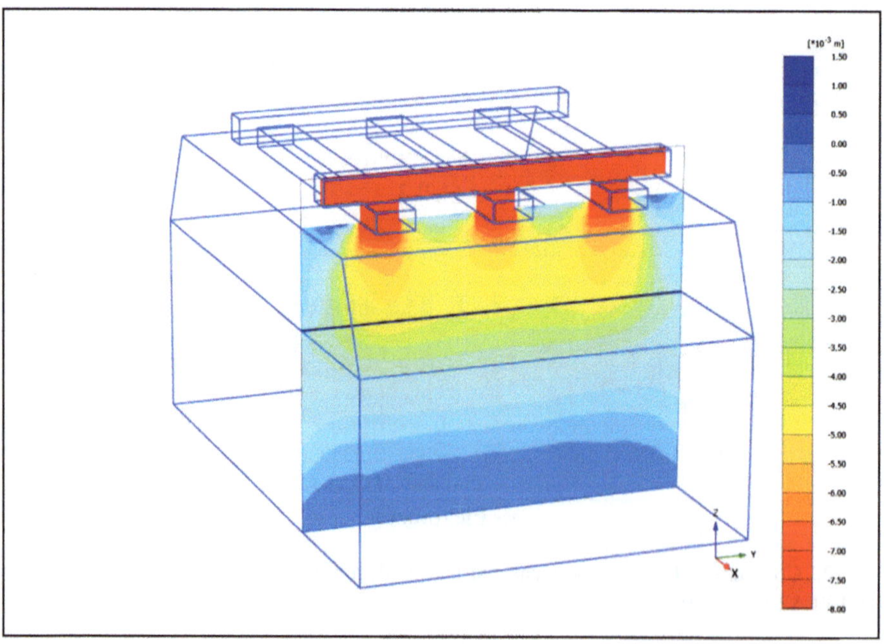

Fig. 24. Displacement in z direction, vertical section at the end of phase three for the test model T30 A2 f1 NL1-0.25.

The reinforcement mechanism of a lateral restraint, or shear-resisting interface, develops through interaction of the ballast layer with the geogrid layer (or layers) contained in or at the bottom of the ballast. Track loads applied to the ballast surface create a lateral spreading motion of the clay layer. Tensile lateral strains are created in the base below the applied load as the material moves down and out away from the load. Lateral movement of the ballast allow for vertical strains to develop, leading to permanent deformation in the wheel path.

The stresses at cell (2) location almost for all tests are starting at a certain level and still at this level for a period, and then the stresses increased rapidly for the next period to reach a maximum magnitude and then decreases for the other period and increases … and so on. This case is likely to occur in cell 2 which is located directly under the loading effect, because of the interlocking between geogrid and ballast, which affects the confinement of the ballast particles, whereas the transmitted stresses increased by increasing the confinement. In the beginning of the test, the interlocking is little and increased with time because of the ballast particles reorientation and at the same time, the geogrid is not effective until a little strain occurs in the geogrid as any reinforcement material, so at the beginning of the test, the ballast receives less confinement than that after a time of starting the test, so the transmitted stress in the beginning is less than after a period of time. After a period of time, there will be degradation in the interlocking of ballast particles with geogrid which decreases the transmitted load because this cell is located in the ballast exposed directly to the action of dynamic load and the cell is close to sloped end of the ballast.

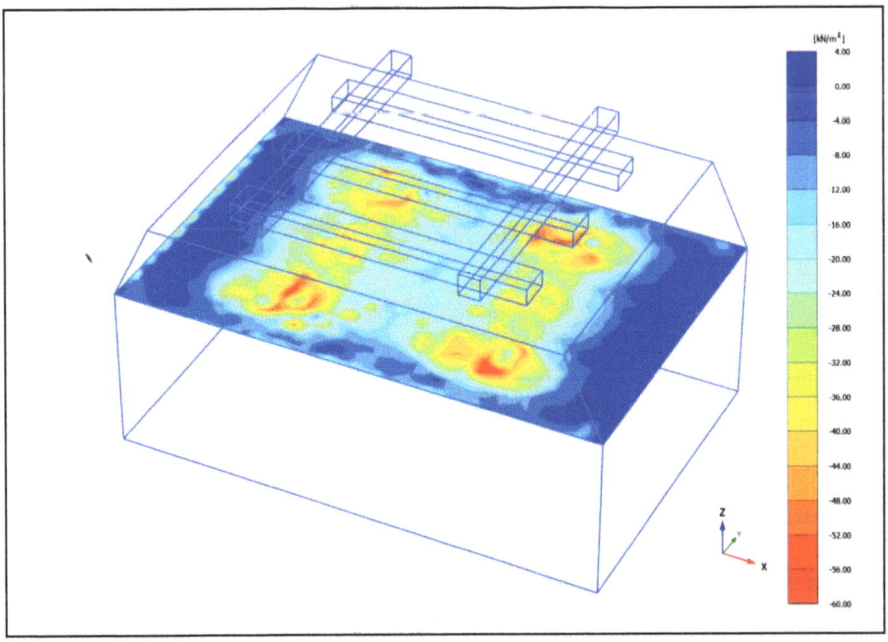

Fig. 25. Total stresses (kPa) in z direction, horizontal section, at the end of phase three for the test model T30 A2 f1 NL1-0.25.

Nareeman and Fattah (2012) found that the shear stress increased for soil reinforced by horizontal geonet layer, decreased. This is because the geonet layer works as a reinforcement layer that strengthens the soil and tends to increase shear strength of the soil. It can be seen that booth compression and dilation of the soil are decreased by adding reinforcement layer.

It is noticed that increasing the ballast layer thickness from 20 cm to 30 cm leads to decrease the settlement by about 50%. This ascertains the efficiency of ballast in spreading the waves induced by the track.

Figure 23 shows contour lines of the total vertical stress along a vertical section blow the center of the rail, while Fig. 24 presents contour lines of the vertical displacement along that section for the test model T30 A2 f1 NL1-0.25.

Figure 25 shows contour lines of the total vertical stress along a horizontal section at clay layer surface, while Fig. 26 presents contour lines of the vertical displacement along that section for the test model T30 A2 f1 NL1-0.25.

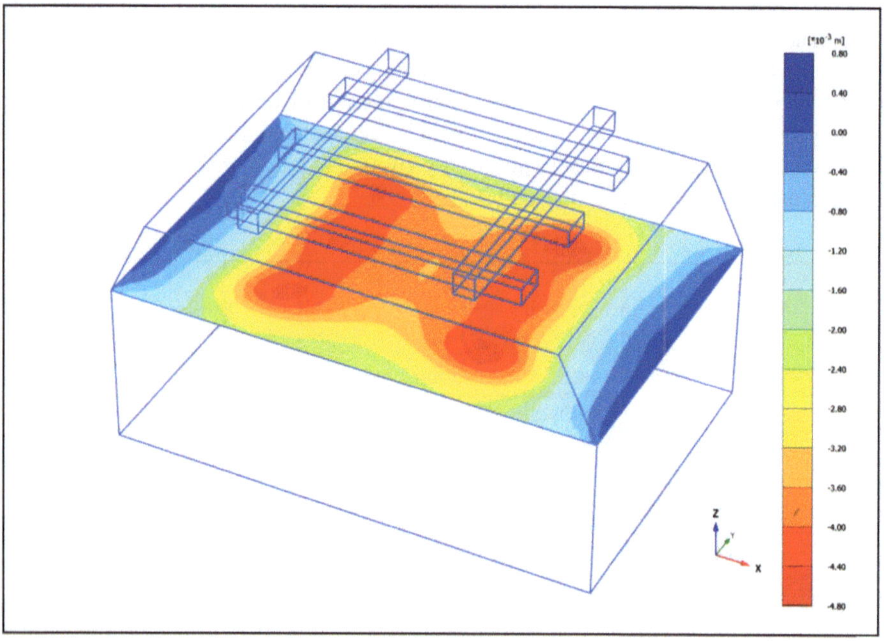

Fig. 26. Displacement in z direction, horizontal section at the end of phase three for the test model T30 A2 f1 NL1-0.25.

3 Conclusions

1. Both experimental and numerical results have the same behavior. For unreinforced case, it was observed that the experimental results at the beginning show higher readings than the numerical ones, but after a number of cycles ranging from 750 to 2500 cycles, the numerical results show higher readings.

2. The developed pressure and pore water pressure for experimental tests have higher readings and a wider range of readings than the theoretical ones.
3. There is a little increase in the induced settlement when the load amplitude increased from 0.5 to 1 ton but it is higher when the load amplitude increased to 2 ton, the increasing amount in settlement depends on the geogrid existence and the other studied parameters. Increasing the ballast layer thickness from 20 cm to 30 cm leads to decrease the settlement by about 50%. This ascertains the efficiency of ballast in spreading the waves induced by the track. The effect of load frequency on the settlement ratio is almost constant after 500 cycles. In general, for reinforced cases, the effect of load frequency on the settlement ratio is very small ranging between 0.5–2% compared with the unreinforced case.
4. The stresses at cell (2), located directly under the loading effect, almost for all tests are starting at a certain level and still at this level for a period, and then the stresses increased rapidly for the next period to reach a maximum magnitude and then decreases for the other period and increases … and so on. This case is likely to occur in cell 2 because of the interlocking between geogrid and ballast, which affects the confinement of the ballast particles, whereas the transmitted stresses increased by increasing the confinement.

References

ASTM D 422-00: Standard Test Method for Particle Size Analysis of Soils. American Society for Testing and Materials

ASTM D 4318: Standard Test Method for Liquid Limit, Plastic Limit, and Plasticity Index of Soil, American Society for Testing and Materials (ASTM)

ASTM D 854: Standard Test Method for Specific Gravity, American Society for Testing and Materials (ASTM)

Awoleye, E.O.A.: Ballast type - ballast life predictions. Derby, British Rail Research LR CES 122, October 1993 (1993)

Heidari, M., El Naggar, M.H.: Using reinforced soil systems in hammer foundations. In: Proceedings of the Institution of Civil Engineers, Ground Improvement, vol. 163, no. GI2, pp. 121–132 (2010)

Indraratna, B., Shahin, M.A., Rujikiatkamjiron, C., Christe, D.: Stabilization of ballasted rail tracks and underlying soft formation soils with geosynthetic grids and drains. University of Wollongong, Australia (2006)

Koerner, G.R., Koerner, R.M.: In-situ temperature monitoring of geomembranes. In: Proceeding GRI-18 Conference at GeoFrontiers, Austin, TX, 6 p. (2005)

Kwan C.C.J.: Geogrid reinforcement of railway ballast. Ph.D. thesis, University of Nottingham, UK (2006)

Leshchinsky, B., Ling, H.I.: Numerical modeling of behavior of railway ballasted structure with geocell confinement. Geotext. Geomembr. J. **36**, 33–43 (2013)

Li, D., Hyslip, J., Sussmann, T., Chrismer, S.: Railway Geotechnics. E-Book. Taylor and Francis Group, LLC, Boca Raton (2016)

Nareeman, B.J., Fattah, M.Y.: Effect of soil reinforcement on shear strength and settlement of cohesive-friction soil. Int. J. Geomate **3**(1 SI. 5), 308–313 (2012). Geotec, Construction Material and Environmental, Japan

Nguyen, K., Goicolea, J.M., Galbadón, F.: Dynamic Effect of High Speed Railway Traffic Loads on the Ballast Track Settlement. Group of Computational Mechanics School of Civil Engineering, Technical University of Madrid (2011)

Profillidis, V.A.: Railway Engineering. Ashgate Publishing Limited, Aldershot (2000)

Raymond, G.P., Bathurst, R.J.: Test results on exhumed railway track geotextiles. In: Proceedings of the 4th International Conference on Geotextiles, Geomembranes and Related Products, The Hague, the Netherlands, vol. 1, pp. 197–202 (1990)

Sowmiya, L.S., Shahu, J.T., Gupta, K.K.: Railway tracks on clayey subgrades reinforced with geosynthetics. In: Proceedings of Indian Geotechnical Conference, 15–17 December 2011, Kochi, pp. 529–532 (2011)

Tutumluer, E., Huang, H., Hashash, Y.M.A., Ghaboussi, J.: Discrete element modeling of railroad ballast settlement. In: Proceedings of the AREMA Annual Conference, 9–12 September 2007, Chicago, IL (2007)

Tutumluer, E., Qian, Y., Hashash, Y.M.A., Ghaboussi, J., Davis, D.F.D.: Discrete element modelling of ballasted track deformation behavior. Int. J. Rail Transp. **1**(1), 57–73 (2014). http://dx.doi.org/10.1080/23248378.2013.788361

Wayne, M., Fraser, I., Reall, B., Kwon, J.: Performance verification of a geogrid mechanically stabilized layer. In: The 18th International Conference on Soil Mechanics and Geotechnical Engineering, Paris, pp. 1381–1384 (2013)

Assessment of Turnout-Related Derailments by Various Causes

Serdar Dindar[1(✉)] and Sakdirat Kaewunruen[2]

[1] Civil Engineering, Birmingham University, Birmingham, UK
SXD319@bham.ac.uk
[2] Railway and Civil Engineering, Birmingham University, Birmingham, UK

Abstract. Train derailments can mainly result in not only financial losses in the form of damaged rolling stock and infrastructure, but also more importantly in causalities and operational shut-down. Therefore, it is crucial for the railway industry to sustain a reliable and efficient operation and eliminate safety concerns. Analysis of accidents caused by train derailment is highlighted as one of the most crucial steps in the risk management chain. Considering various operational environment, the analysis enables reduction in the occurrence of derailment and prevents derailments in the most cost-efficient manner, as a variety of different causes and their frequency and severity are determined. In this paper, the methodology includes gathering and analysis of information on major derailments occurring at the turnouts from the UK. The causes have been categorised and then prioritised in accordance with the proportion of train derailments occurring within each category. The research objective is to determine the proportion of train derailments at turnouts and provide a starting point for the detailed analysis. In short, the aim of the paper is firstly to understand which factors under which circumstances pose the greatest risk at turnouts, secondly to quantitatively evaluate the relationship between derailment severity and frequency and thirdly, to determine the characteristics that cause major derailment and compare results with the characteristics in mainlines. The review analyses train derailments in UK over the last 15 years and demonstrates that the overall number of derailments have declined gradually and most derailments have occurred in yards. The dominant causes are identified as operational failures and component faults during track-train interaction, and track geometry problems are also seen to have a significant impact on derailment at turnouts, particularly in shunting yards. Furthermore, literature-based recommendations are used to address the issues arising from risks. The research outcomes are expected to aid the rail industry in developing, evaluating, prioritizing and gaining different perspective of derailment at turnouts to efficiently improve transportation safety and also to open a new gate to better understand the existing risk on railway turnouts.

1 Introduction

Train derailments are one of the major undesirable events to be considered in railway safety system and potentially pose serious hazards to human health and safety. A derailment might occur through many reasons, including operational error, environmental

J. Pombo and G. Jing (eds.), *Recent Developments in Railway Track and Transportation Engineering*, Sustainable Civil Infrastructures, DOI 10.1007/978-3-319-61627-8_2

impacts, collision, various mechanical failures, etc. Additionally, it is notoriously familiar phenomenon likely to occur on all railway infrastructures, such as turnouts, track rails, etc. As a result, any risk mitigation of the occurrence of a derailment requires a broad knowledge.

In recent years, domestic scholars have conducted several researches related to the statistical assessment of train derailments. Liu et al. (2012) firstly analysed of train derailments by accident cause. This research proposed a model assessing derailment risk by accident cause and FRA track class. The same research group (Liu et al. 2011) subsequently indicated the statistical results of major train derailments and their effect on accident rates. However, the research only dealt with railway freight train safety, excluding other types, such as passenger trains and engineering trams. It is worth noting that there is no significant conclusion of any relationship between causes and turnout-related derailments.

On the other hand, it has been revealed that an effective and sustainable risk management of derailments at turnouts might be achieved through individual investigation of all railway infrastructures (Dindar and Kaewunruen 2016). As a consequence, turnouts, as one of the most complex railway engineering systems, should be taken individually into account, considering a variety of factors, e.g. type of equipment, line, along with immediate cause, causal and contributory factors of accidents investigated, as reported by a reliable and unbiased organisation.

Turnout-related derailments collectively cause almost the half of all train derailments in the UK (Ishak et al. 2016). Understanding what types of causes contribute to derailments at turnouts is highly likely to provide the fundamentals for preventing such accidents as well as developing cost-effective maintenance strategies. This also results in reducing railway transportation risk, allowing the railway operators to sustain a smooth operation.

The study illustrates the statistical results of derailments at turnouts, and then evaluates and discusses the results to achieve the initial step in a systematic process of quantitative risk analysis and risk management for all operational conditions.

2 Turnouts

In railway engineering, the term turnout, as shown in Fig. 1, is used to describe a mechanical installation by means of which flanged vehicles are able to be diverted from one track to another. Although turnouts have a close similarity in their design and method of installation, the engineering process in building their constituents may be vastly different, giving rise to their complexity.

The various constituents of a turnout are shown in Fig. 1. A brief definition of the each is given below along with a technical explanation of the significant details

Stock rail: the main rail of the track to which the tongue rails are closely fit.

Points or switch: Steel blocks formed by the combination of a pair of tongue and stock rails with the necessary connections and fittings.

Crossing: an arrangement of rails introduced at the junction where two rails cross each other to allow the wheel flange of a vehicle to be moved from one track to another.

Fig. 1. Typical components of a turnout

A switch motor: an electric, hydraulic or pneumatic mechanism used to align the switch with one of the possible routes.

Stretcher Bar: a steel bar used to keep the switch rails in the correct position under the passage of a rolling stock.

Sleepers: They are generally laid perpendicular to the rails and transfer loads from the rails to the track ballast and subgrade. Their further benefits are to hold the rails upright and keep them spaced to the correct gauge.

Closure rail: the piece of fixed rail between the points and frog of a turnout to provide the passage of a rolling stock from switches to crossings.

Heel block: a unit providing a splice with the contiguous closure rail and a location for the switch point rail.

Check rail: a short piece of rail placed alongside the stock rail opposite the frog to ensure that the appropriate flange way through the frog is followed by wheels

Turnouts are operated in a variety of ways in accordance with type of turnouts, type of line and operational environment. Switch rails can be moved by a human operator or a switch motor. Although some switches, in particularly light and tram lines, are still controlled by humans, most, situated on high speed, new and heavy lines, is preferred to have a remotely controlled electric or hydraulic or pneumatic motor.

3 Data Analysis

To achieve generalisability of the paper and targeting identification of specific risk levels of railway turnout systems, it is essential to argue how reliable, sufficient and accurate data are used in this research and to what degree data give an insight into railway safety industry. This section offers a response to such concerns.

3.1 Data Source

The research is limited to investigation of UK-based accident causes. Therefore, there is a variety of British sources presented for this analysis. The Rail Accident Investigation Branch (RAIB) is one such and is a British government agency that conducts enquiry into various types of train accidents in the UK. Created in 2005, the agency has official authority to investigate accidents that cause death, serious injury or extensive damage to infrastructures. It more specifically focuses on accidents and incidents on mainline railways, metros, tramways and heritage railways throughout the UK, operating as an independent body.

Composed of a total of 43 persons, their inspecting team comprises experienced rail or investigation specialists. Accident reports are officially reliable and controlled by the Secretary of State for Transport. The accident reports of RAIB include as follows;

- the immediate cause, causal and contributory factors of railway incidents and accidents
- identifying risks which might result in a similar accident or make an accident worse and making recommendations to prevent reoccurrence
- detailed information about how railway accidents happen

All types of accidents occurring at turnouts in the UK since 2006 have been investigated.

3.2 Data Categorisation

Rail Accident Investigation Branch reports are formed from a wide range of safety events, including accidents, incidents and near misses, and extend beyond the operation-focussed requirements of Railway Group Standard GO/RT3119 and Accident and Incident Investigation Guidance (RSSB 2016). It is found that RAIB has published over 130 derailment reports related to accidents occurring on heavy, light, metro and heritage lines. It is determined that there are 45 turnout-related accidents occurring on such all lines. These reports have also been investigated as to whether these are accidents with a monetary threshold of damage to infrastructure and rolling stock. Lastly, it is decided that 44 derailments meet the principles by which risk assessment may be judged to be initiated.

A method is developed to categorise immediate, causal, contributory and underlying factors of accidents (Marshall and Healey 2008). It is observed that RAIB and RSSB largely work from a model and a similar methodology, modified to take account specifically turnout-related problems in the context of technical rail categorisation, is used for this study. The classification is represented in Fig. 2. The main difference between the modified model and the model currently in use is that the new model proposes the separation of infrastructure component failures from non-infrastructure component failures, e.g. bogie faults, wheel defects, etc. As it is noticed that there have been many human-based failures, the failures were handled separately without adding to any major category.

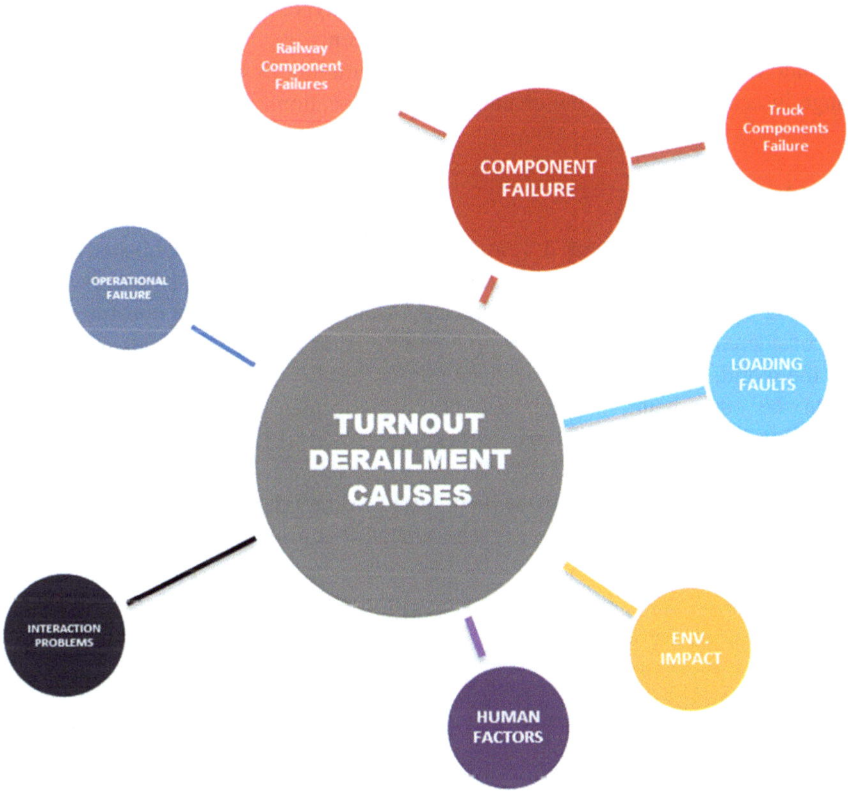

Fig. 2. Representation of proposed framework for categorisation

Briefly, all major categories, including operational failures, environmental impacts, interaction problems, and human factors, are given along with their subcategories in Table 1.

Table 1. Subcategories of all major accident nodes

Infrastructure	Interaction	Environmental	Operational	Loading	Human
Switch	Obstruction	Rain	Signal	Malicious	Speed
Points	W/R	Snow	Use of switch		Br.Rules
Stretcher Bar	Flange Climb	Mud	Brake operation		Fatigue
Geometry	Hunting	Wind	Train handling		Vandalism
Ballast					

The paper also attempts to categorise railway type, e.g. heavy, light, metro and heritage lines. Furthermore, the location of accidents, e.g. yard, siding, plain line, is taken into account. This could be useful for determining to what degree the risk levels change in accordance with the characteristics of location.

3.3 Limitation

To understand better the work developed here, this is the section established for further studies in the field. The authors obtained data from only RAIB as data source, without comparing the results with another database. Although this external constraint present, it is considered that the results can probably be addressed to the issues in general all over the world.

On the other hand, the database has been formed only of UK-based reports. The UK railway industry is one of leading expert regarding minimising railway accidents, e.g. derailment, collusion. Hence, where any developing country are studied, the results of this study are likely to be different. Although this study are designed well for future risk method and reliability assessment, it is limited to advanced countries in railway sector.

4 Results

4.1 Derailments by Railway Type

Four types of railway types are recorded in the RAIB reports, namely heavy, light, metro and heritage rails. Different operational functions might be executed through these different railway types and, consequently, are expected to lead to different causes, accident types and consequences.

Heavy rail refers to conventional railways forming part of the national network, including high-speed rail, freight, intercity, commuter and rural services. RAIB and RSSB use the term light rail to express high capacity lines on which urban public transport, using rolling stock similar to a tramway, are used and are generally established on an exclusive right-of-way. Metro might be seen as underground light rail, but cannot be accessed by pedestrians or other vehicles of any sort. As for heritage rails, the means that volunteers or non-profit organisations take over or re-open railway lines which were once run as commercial railways.

Figure 3 shows accident frequency and wagon derailment by railway type, considering all reported cases in the UK from 2006 to 2016. It provides evidence that the distribution of accident types varied in accordance with railway type. Train derailment on heavy lines is seen to dominate the overall accidents. Over the last ten years, 35 significant derailment cases, accounting for 77% of all, have occurred on such lines. The most common railway type, it is found to be followed by light rail, metro and heritage railways.

Risk comprises two fundamentals; likelihood and consequences/severity. As the report does not include cost of accidents, e.g. track or rolling stock maintenance expenditure, it is assumed that accident severity is determined in this research by the

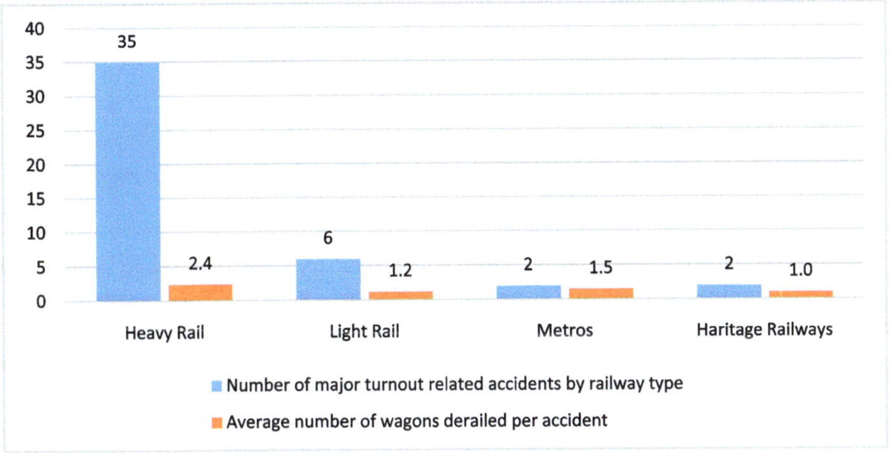

Fig. 3. The accidental characteristics of turnouts by railway type

number of wagons derailed per accident. Similarly, train derailments on heavy lines, an average number of 2.4 per accidents, have had a greater average accident severity than other railway types, namely metro, light rail, and heritage rail, at 1.5, 1.2., 1, respectively.

4.2 Derailments by Railway Location

Three types of railway line location are recorded in the RAIB database, namely main, siding, yard and industry line. As with railway type, these are necessary for achieving different operational functions and, as a result, have different associated accident causes and consequences. Considering the many variants in railway terminology, main line, in this research, refers to a stretch of railway track that is away from yards and sidings. Yards are a complex series of railway tracks used for loading/unloading, sorting or storing railway trucks, while sidings, in British terminology, are tracks where railway vehicles can be left, i.e. are not serving as an operating train for the time being. There are some lines that are not for public transportation, but for serving a particular industrial, logistic or military purpose. These lines are addressed as Industry lines.

RAIB turnout-related accident reports covering the period 2006 to 2016 were compiled to illustrate the number of major accidents by railway type, the average number of wagons derailed per accident and average speed of derailed vehicles (Table 2). It is worth noting that accidents resulted by railway locations include both passenger and freight trains.

These three statistical categories were dominated by accidents on main lines, as expected due to their high traffic density. On the other hand, although yards are equipped with small pneumatic, hydraulic or spring-driven braking retarders to slow train speed, given the British speed regulations, and that these are very common railway areas, there seems to be a relatively greater number of turnout-related accidents

Table 2. Statistical results by railway location

	Main	Yard	Siding	Industry
Number of major turnout related accidents by railway type	28	10	7	1
Average number of wagons derailed per accident	2.4	1.6	1.5	2.0
Average speed of derailed vehicles (mph)	14.9	10.5	5.6	N/A

than expected. This will be discussed later. As for industry lines, these have been added to make the findings usable for US researchers. However, as only one accident has so far been recorded, it is quite difficult to make any conclusion on the line.

Although serious accidents are seen likely to occur on the yards and siding areas of a railway line, main line operation needs to be considered carefully to maintain a smooth operation of the entire railway network, because heavy trains on this part travel at high frequency and higher speed. Another feature of the table is shows the explicit relationship between speed and number of accidents/average number of wagons derailed per accident. The higher the speed at which the rolling stock travels, the more accidents and derailed wagons are highly likely to occur. It could be considered that the rolling stock on main lines are of greater mass than those on other lines. The greater mass and speed signifies that the rail-wheel interaction forces and potential impact in regard to casualties, property and environmental damage, are expected to be all correspondingly greater. This could be an explanation in the logarithmic rise in the number of derailed wagons against rolling stock speed.

4.3 Derailments by Accident Cause

Derailments are the most common type of train accident type in the United Kingdom. The rail industry and government-based organisations concentrate on preventing them to provide a high standard of railway operation and eliminate potential safety concerns. Several researches have focused on train derailment and causes, using often the Federal Railroad Administration (FRA) database. The administration reports a large number of accidents covering almost the last three decades in the United States, and these reports, each of which is of one or two pages, are as detailed as RAIB reports. On the other hand, the U.S. has a length of nearly 150,000 miles and over 1.2 billion tonne miles of freight rail usage, which is remarkably higher than the statistics in the UK (about 20,000 miles, 23 tonne miles of freight rail usage, respectively) (NS 2015). Considering the relation between the figures and accidents, the study has dealt with a considerable low number of accidents. However, reports present much more detailed information on immediate causes, causal factors and contributory factors. The latter two have not been included in FRA reports.

Most previous studies have concentrated on main line, yard and sidings derailments; however, in this research, the paper only focuses on turnout-related derailments which demands for elaborated explanation on how a rolling stock is derailed at turnouts. Each accident has been investigated to rank immediate causes, causal factors and contributory factors of derailment at turnouts.

In this study, immediate causes are defined as substandard acts or various conditions that lead directly to the derailment. In the event of avoiding or eliminating any immediate causes, associated accidents would be prevented from happening. Likewise, causal factors too are any condition, event or behaviour that was necessary for the occurrence of derailment and can be anyone of these factors. However, some conditions, events or behaviours might affect or sustain the occurrence, or exacerbate the derailment (contributory factors). Eliminating one or more of these factors would not prevent derailment, but their presence makes it more likely.

Figure 4 illustrates the percentages, by each immediate cause, leading to derailments at turnouts in the UK. Each has several subcategories, as shown in Table 1. Some accidents were reported with two or more immediate causes and those were added each by each into analysis in the figures and tables in this section.

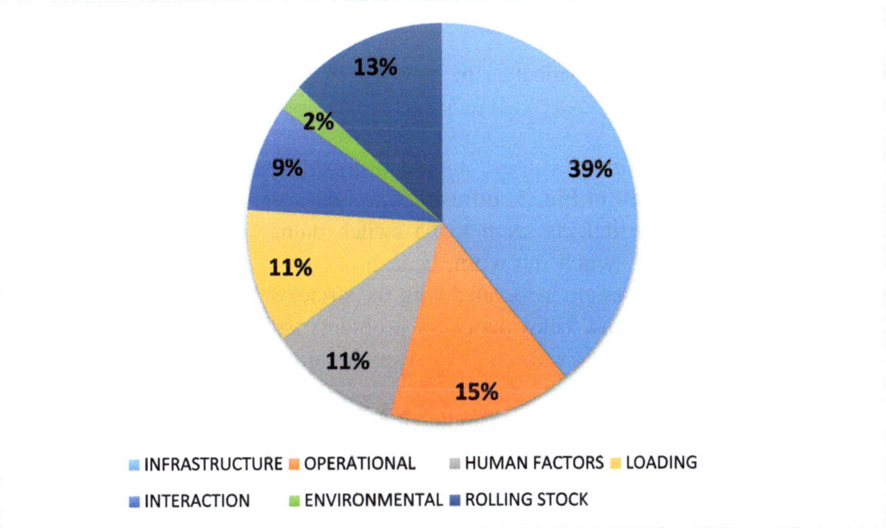

Fig. 4. Distribution of derailment frequency by various causes

Infrastructure associated problems, e.g. broken rail and various turnout component failures, have contributed to the majority of derailments. This could be expected due to the fact of the extreme force and potential impact in regard to turnout component damage as a result of discontinues on the rail line by railway turnout design itself. The results statistically express how vulnerable turnout components are to the forces of every passing rolling stock. Although the infrastructure failures account for the vast majority of derailment, operational and human factors are almost the same contribution, comprising 36% of accident rates combined. Of the two, 15% were caused by operational failures, while another 11% were caused by human factors. The rest were caused, in order, by loading faults (11%), interaction problems (9%) and environment (13%). As this study is limited by excluding rolling stock-based faults, which demand for a different expertise area, there is no discussion in regards to this.

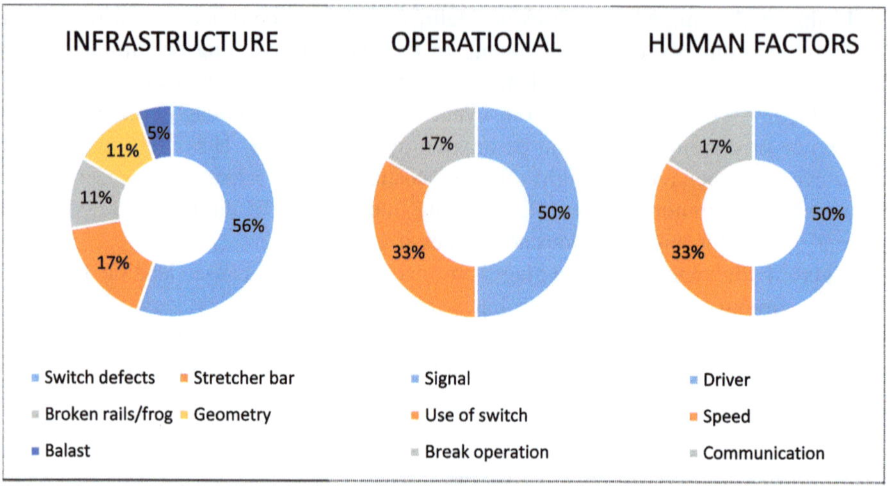

Fig. 5. Derailment frequency distribution by subcategory-indexed reasons of the top three factors

The top three categorised causes, namely infrastructure, operation and human factors, are given in detail in Fig. 5. Infrastructure-related causes mainly involve various switch defects. The defects seem to be switch damaged or out of adjustment, switch point gapped and switch rod worn, bent, broken, or disconnected. The second infrastructure-related causes are associated with the stretcher bar not working properly. The rest are broken rails and frogs, geometry problems and ballast.

Signal problems account for the half of operational-caused derailments. As such, causes include flagging, improper or failure to flag, or the restrictive indication of a block or interlocking signal or wrong indication of signal. Such problems were followed by use of switch and brake operational failures.

Lastly, human-caused derailments are mostly traced back to various driver-centred failures, i.e. failure in not reacting to the 'points not correctly set' indication on the signal or not immediately observing what was displayed by the signal's route indicator. Those are followed by disobeying the permitted speed and communication errors, such as improper radio communication or failure to comply or give/receive.

Aside from immediate causes, causal factors are very important in determining the cause or causes of the derailments so as to prevent further turnout-related accidents of a similar kind. Table 3 shows such groups, indicating how responsible each is for derailments. It is evident that derailments at turnouts are mostly a man-made crisis, with maintenance faults, human error and signalling problems accounting for over 50% of accidents. Among these causal groups, maintenance problems appear to be of most interest and need to be reviewed in current maintenance regimes due to the unacceptable frequency of occurrence. As a common result of the complexity of its geometry, with extreme unique internal/external forces, and vulnerability to environmental conditions, it is recommended that each individual turnout should be taken into account for maintenance strategies. The railway industry will obtain benefit from this in eliminating risk factors associated with derailments (Dindar and Kaewunruen, Investigation of

Table 3. Statistical results of causal factors

Causal Groups	Description	Number of derailments	Percentage
Maintenance	Installation/regime problems, insufficient inspection, undetected bearing/rail welding.	14	32.6%
Environment	Extreme rain, icing/snow on track, high temperature	7	16.3%
Human factors	Vandalism (e.g. by someone placing a stone), slow response to signs or communication	6	13.6%
Track geometry	Significant twist fault, widening has occurred, track gauge failures	5	11.4%
Rolling stock	Degraded bogie/wheel	5	11.4%
Sign problems	Wrong settlement of points set indicator/related signs, blocked signs	2	4.5%
Malicious	Train lateral/horizontal forces, vulnerable components, design problem	5	9.1%

Risk-based Maintenance Strategies 2016). Among outer factors, about 16% of derailments were related directly to environmental conditions as causal factors. A recent study (Dindar and Kaewunruen, Natural hazard risks on railway turnout systems 2016) has already provided the link between environmental factors, e.g. hot/cold temperature, and derailments at turnouts. Stretcher bars and turnout geometry have also been found to be vulnerable.

Table 4 gives information about the distribution of contributory groups analysing the 46 accident reports. The table is prepared to provide insight into how the likelihood or severity of derailments at turnouts escalates. Although some group names in Table 4 are the same as those in the previous table, the contents of each are different. For instance, environment is death as a contributory and causal factor. Extreme rain or icing/snow on track or high temperature is thought to directly cause derailment in Table 3, whereas moderate environmental patterns, such as low visibility due to fog, are considered to promote derailment accidents.

One of the contributor groups, maintenance, was the most frequently reported category, involved in 20% of all derailment reported by RAIB. This group is mainly built on the concerns described in the Table. It is seen that a low frequency of maintenance or an inappropriate inspection or the lack of independent inspection is often reported.

On the other hand, sign/signal-related factors were the second most frequently reported group, involved in 18% of all reported accidents. The group mainly concerns the close proximity of points to signal and the absence of an illuminated PSI (points set indicator). Only two accidents are reported by unseen railway lights (due to being blocked by a tree) and small size of railway sign.

Components as a contributor group were the next most frequently reported category, associated with 16% of the accidents. It is mostly reported that, firstly, various rolling stock components, such as a bogie, are not detected to be degraded; secondly,

Table 4. Statistical results of contributor groups

Contributory groups	Description	Number of derailments	Percentage
Maintenance	Lack of detailed inspections, low maintenance frequency, inspection error, inappropriate maintenance plan, etc.	10	20%
Sign/Signal	Location problems, size problems, absence of various aimed signal boxes, etc.	9	18%
Component	Stiffness-related failures, not-fitted-in-use, somewhat out of standard components, etc.	8	16%
Environment	Fog, rain, warm, cold weather patterns, etc.	7	14%
Human errors	Fatigue, mediating drug use, etc.	7	14%
Geometry	The design of the points, the presence of voids, etc.	5	10%
Operational	The failure of the control room, inaccuracy of appointments in inspection, etc.	4	8%

the bolt retaining plates on the field-sides of both switch rails have not been fitted properly; and, thirdly, the lack of support given by the stock rail to the switch because of loose fastenings.

The other groups in the Table, environment, human error, geometry, and operational, are responsible for some contribution to almost half of the accidents. The details of each are described in the Table.

5 Conclusion

A risk analysis often begins with identification of risks and their various contributing factors. The study determined risk groups and prioritised them to establish a most-to-least critical importance ranking. As a result, the outcomes of the study provide insight to risk analysis/management of railway turnout systems whereby their immediate, causal and contributory factors need to manage or mitigate the realisation of the high probability/high consequence of risk events.

Train derailments at turnouts are observed to account for the majority of accidents that occur on heavy lines. Although metro lines (underground) have as high a density of traffic as heavy lines, average derailments on metro lines are at a considerably lower rate. Light rail, as an urban transportation system at ground level, is also determined to be responsible for many turnout-related derailments and identified to be very vulnerable to such accidents.

It is illustrated that turnout location, e.g. siding and yard, is examined to determine its strong correlation with derailment rate. Rolling stocks on main line are at more risk than those on siding, yard or industrial lines. Moreover, the greater the permitted speed over turnouts, the larger number of accidents it yields. However, the relationship between speed and derailment is complex, and demands further work. Some accident causes are identified to show a strong relationship, but others do not.

The main statistical information on derailments at turnouts is illustrated in figures prepared through official accident reports published by the Rail Accident Investigation Branch. The top causes for derailments are infrastructure, operational and human factors, respectively. The proportion of infrastructure associated with various turnout component failures is the highest. The most failing component at turnouts is identified as switch. Stretcher-related failures are the second most common failure type. Aside from component failures, operational and human factors seem to contribute significantly to derailments.

The most significant causal factor is maintenance, accounting for one-fourth of all. The major reason is reported as an undetected failing element or, in other words, problems in maintenance procedures. This is followed by environmental factors and human factors. Turnouts seems to very vulnerable to cold/hot weather, causing various geometry problems, and to driver faults such as communication.

The paper also dealt with contributory factors. Maintenance is the most common type and, among the reasons, low rate maintenance frequency is reported; therefore, increasing the frequency is recommended by British derailment experts.

Acknowledgements. The first author is grateful to Turkish Ministry of Education for his postgraduate research scholarship. The authors wish to thank European Commission for the financial support of this research via H2020-RISE Grant No 691135 'RISEN: Railway Infrastructure Systems Engineering Network'.

References

Dindar, S., Kaewunruen, S.: Identification of appropriate risk analysis techniques for railway turnout systems. J. Risk Res. 1–22 (2016). doi:10.1080/13669877.2016.1264452

Dindar, S., Kaewunruen, S.: Investigation of risk-based maintenance strategies. In: ART-2016, The Korean Society for Railway, Juju, Korea (2016)

Dindar, S., Kaewunruen, S.: Natural hazard risks on railway turnout systems. Procedia Eng. **161**, 1254–1259 (2016). doi:10.1016/j.proeng.2016.08.561

Ishak, M.F., Dindar, S., Kaewunruen, S.: Safety-based maintenance for geometry restoration of railway turnout systems in various operational environments. In: Proceedings of The 21st National Convention on Civil Engineering. Songkhla, Thailand (2016). http://www.ncce21.org/home/index.php?l=th

Liu, X., Barkan, C., Saat, M.: Analysis of derailments by accident cause: evaluating railroad track upgrades to reduce transportation risk. Transp. Res. Rec. **2261**, 178–185 (2011). doi:10.3141/2261-21

Marshall, M., Healey, N.: Review of RAIB investigations and recommendations 2007; RSU/08/12. lONDON: ORR (2008)

NS: Transport Statistics Great Britain 2014. Department for Transport, London (2015)

RSSB: Guidance on Hazard Identification and Classification: GE/GN8642. Rail Safety and Standards Board Limited, London (2016)

Liu, X., Saat, R., Barkan, C.P.: Analysis of causes of major train derailment and their effect on accident rates. Trans. Res. Board **2289**(1), 154–163 (2012). doi:10.3141/2289-20

Performance Level of Road Geometric Design Based on Motorcycle – Cars Linear Speed Profile

Joewono Prasetijo[1(✉)], Guohui Zhang[2], Zaffan Farhana Zainal[1], Wan Zahidah Musa[1], and Nickholas Anting Anak Guntor[1]

[1] Faculty of Civil and Environmental Engineering, Smart Driving Research Center (SDRC), Universiti Tun Hussein Onn Malaysia, 86400 Parit Raja, Johor, Malaysia
joewono@edu.uthm.my

[2] Department of Civil and Environmental Engineering, University of Hawaii at Manoa, 2540 Dole Street, Honolulu, HI 96822, USA

Abstract. Malaysia is one of the developing countries that face the highest rates of road accidents in Asia. The most common accident happens between motorcyclist and cars. A motorcyclist is 17 times more dangerous than a passenger car. Analysis shows that the three main types of accidents in Malaysia are: collisions with passenger cars, collisions with other motorcycles, and single-motorcycle accidents. Road accident rates continue to rise in Malaysia, due to the lack of geometric design consistency on roads, wherein drivers make errors due to the geometric features of roads. Several prediction models are developed in this study for estimating operating speed on tangent sections, depending on the characteristics of the tangent, and on continuous speed profiles. Geometric factors such as access point, tangent length, curve length, shoulder width, and lane width influence operating speed profiles for the development of tangents and curves for motorcycle and cars. The study was conducted at F0050 of Johor federal roads. Furthermore, the continuous speed profile data were used to develop road design consistency profiles for cars and motorcycles (IC). The development of the consistency model is based on these parameters: the bounded area between the profile and the average speed and standard deviation of speed along a segment. The highest areas of an accident at km24, which are justified based on the design consistency of the area and on the integrated design consistency model between car and motorcycle, are of a poor design. The poorer the design consistency, the higher the potential for accidents. The study also shows that the model are well compare to the real accidents data of cars and motorcycles.

1 Introduction

Every year, nearly one million people are killed, three million are severely disabled for life, and thirty million are injured in road accidents, according to WHO. As we approach 2020, road accidents are predicted to be the third most important in contributing to the global burden of disease and injury. Road accidents in Malaysia are the

J. Pombo and G. Jing (eds.), *Recent Developments in Railway Track and Transportation Engineering*, Sustainable Civil Infrastructures, DOI 10.1007/978-3-319-61627-8_3

one of the fearsome issues when it comes to discussing road traffic in the country. Malaysia is one of the developing countries that face the highest rate of road accidents in Asia. About 300,00 accidents are reported in Malaysia every year. with 1% of these being fatal. According to a report from the Royal Malaysia Police, traffic accidents in Malaysia have been increasing over the last three decades with an average rate of 9.7% per annum. With the large volume of traffic on the road, accidents can happen involving various types of vehicles (Abdullah and Zamri 2012). The most common accident occurs between motorcycles and cars. A motorcycle is 17 times more dangerous than a passenger car, and the motorcyclist also faces a higher risk of injury or fatality; the risk of injury is estimated to be 12 higher for motorcyclists than for car passengers (Radin 2005). Analysis shows that the three main types of accidents in Malaysia are: collisions with passenger cars, collisions with other motorcycles, and single-motorcycle accidents. In accidents that occurred in which either the car driver or motorcyclist were not injured, but the injured parties were either passengers or the occupants of the other vehicle, around 97% of motorcyclists were injured or killed in these collisions compared with 50.5% of car drivers. It also stated that at higher speed in motorcycle and car collisions, as opposed to car and car collisions, motorcyclists involved were 95.4% likely to be injured while car drivers were 0.9% likely to be injured (Hung et al. 2006).

Some research states that road accidents happen due to driver fault, mechanical mistakes such as brake failure, burst tyres, etc. However, most road accidents happen because of the road conditions itself (Abdullah and Zamri 2012). The rate of road accidents keeps rising in Malaysia, due to the lack of geometric design consistency on roads, wherein drivers make errors due to the road's geometric features. If the consistency is poor, the potential for accidents is high (Lamm et al. 1999 and Juan de Oña et al. 2012).

1.1 Operating Speed Model (85th Percentile)

Over the past 50 years, several models have been developed to predict operating speed using geometric and traffic characteristics (Castro et al. 2008). Many studies were developed using different data collection devices to estimate the 85th percentile of speed on two-lane rural roads on horizontal curves and tangents for a variety of speed limits and vehicle types. Most of the models developed were based on spot speed data collected at specific locations; the speed measurements used were collected by radar guns, which can result in some inaccuracy because the reading angle is different from zero and drivers may reduce the speed.

Some researchers determine the operating speed profile from the radius of circular curves, deflection angles, the relationship between curve length, etc. In a previous study (Memon et al. 2012), only passenger vehicles operating under free-flow condition were considered. Most prior studies identified the radius of the curve as an influencing parameter in estimating operating speeds. Some studies developed operating speed prediction models based on the continuous operating speed data collected through VBox.

Based on previous studies (Mattar-Habib et al. 2008), several models have been developed for estimating operating speed on tangent sections depending on the characteristics of the tangent and continuous speed profiles, in order to determine the global speed variation along a road segment and to determine a single consistency value for the whole road segment.

1.2 Design Consistency Model

Design consistency is measured by estimating the speed variation along successive geometric elements and evaluating the speed deviation from the average speed in a road section, resulting from geometric features which may cause drivers to make speed errors leading to a higher collision risk (Mattar-Habib et al. 2008). A road with a good level of consistency can result in a driver driving in a stable and careful manner. With a poor level of consistency, however, the result can be high speed along a different path among different drivers, thus causing accidents (Prasetijo et al. 2015). Geometric designs are influenced by the vehicles, drivers, and traffic characteristics. Geometric design elements that influence traffic operations include the number and width of lanes, the presence and width of shoulders and highway medians, and the horizontal and vertical alignment of the highway. The criteria most used to evaluate design consistency are based on evaluations of the operating speed of the 85th percentile speed of vehicles, and are obtained by using operating speed prediction models. Some researchers have extended the concept by using continuous speed profiles (normalized area bounded by speed) to determine the speed variation along a road segment and determining a single consistency value for the whole road segment (Gibreel et al. 1999; Polus and Mattar-Habib 2004; Mattar-Habib et al. 2008). Consistency evaluation methods are based on estimations of the operating speed profile that has been used in other countries including the USA, Switzerland, and Germany. According to Mattar-Habib et al. 2008, the purpose of new methods is to check design consistency. The parameters for this model are the bounded area between the profile and the average speed along a two-lane highway segment. As design consistency increased, crash rates are shown to decrease significantly. The corresponding equations are

$$R_a = \frac{(\sum a_i)}{L} \tag{1}$$

where R_a is relative area (m/sec) measure consistency, $\sum a_i$ is sum of i areas bounded between the speed profile and the average operating speed (m^2/sec), L is entire segment length (m)

$$\sigma = \left\{ \frac{(V_j - V_{avg}}{n} \right\}^{0.5} \tag{2}$$

where σ is standard deviation of operating speeds (km/h), V_j is operating speed along the jth geometric element (tangent/curve) (km/h), V_{avg} is average weighted (by length) operating speed along a segment (km/h), n is number of geometric elements along a segment (km/h)

$$V_{avg} = \frac{\sum V_i L_i}{\sum L_i} \tag{3}$$

where V_{avg} is average weighted (by length) operating speed along a highway segment (km/h), V_i is operating speed (km/h), L_i is length of segment (km).

The model of consistency (C) and the model integrated design consistency is [8]

$$C = 2.808.exp^{(-0.278.R_a.\sigma)} \tag{4}$$

where C is basic consistency of a highway segment, R_a is normalized area bounded by the average speed profile of cars/motorcycle and the average operating speed (m/sec), σ is standard deviation of cars speeds (m/sec)

$$IC = \left[2.808 * exp^{(-0.278 * R_a * \sigma)}\right] * exp^{(-0.01 * A_{CT})} \tag{5}$$

where IC is integrated consistency of road sections, R_a is normalized bounded area by speed profile of cars/motorcycle (m/sec), σ is standard deviation of car speeds/ motorcycle, ACT is normalized bounded area between the speed profile of cars and motorcycle (m/sec)

To determination threshold for design consistency proposed by (Mattar-Habib et al. 2008) are; good design if C > 2 (m/s); acceptable design if 1 < C ≤ 2 (m/s); and poor design if C ≤ 1 (m/s).

2 Data Collection

The purpose of the study is to determine the **P**erformance Level of Road Geometric Designs Based on Motorcycle – Car Linear Speed Profiles. The data collection can be determined based on location, type of vehicle, data speed of vehicles using global positioning systems (GPS), and road features. The study was conducted on Johor Federal Road (F0050), from km21 until km25 located in College Community Nurses Batu Pahat until Taman Manis, Parit Raja, Johor, Malaysia. Figure 1 shows the location km21 until km25.

Speed data collection (primary collection) along the selected segment was carried out by the test driver method during the daytime, under free-flow, and off-peak. The GPS DG-200 was used in this study. GPS was placed int the moving vehiclesused in this study, cars and motorcycles. One hundred test drivers and riders were instructed to drive and ride along the selected segment using the same vehicles. Speed data from signalized intersections were removed at about 300 m to avoid the effect of traffic control devices on vehicle speeds and free-flow speed (Schurr et al. 2002). The GPS device was configured to record the vehicle positions and speeds at one-second time intervals. Two persons, a driver and a GPS recorder, were needed to perform the data

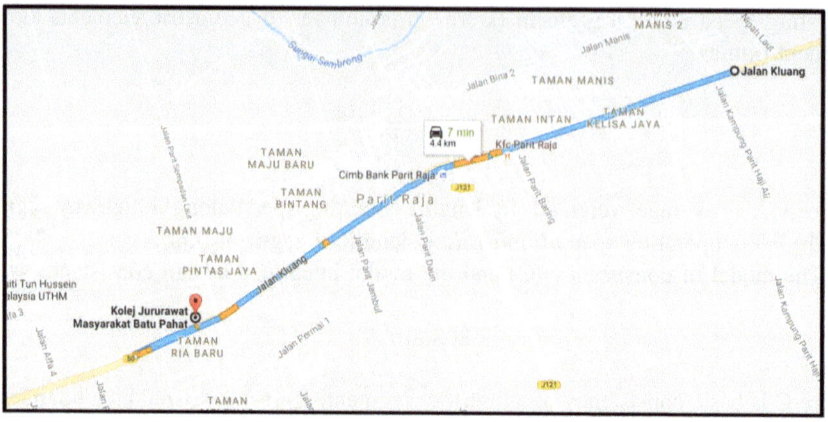

Fig. 1. Location km21–km25 (F0050)

collection and the average vehicle travel speeds along the distance traveled as measured by the GPS at each segment (Prasetijo and Zainal 2016). By using GPS, the continuous speed profile data were using a new analysis method to develop road design consistency profiles for motorcycles and cars and therefore determined the value of the index for the integrated road design consistency (IC).

Secondary collections included data from accident data, traffic data, and road features. The accident data collection was collected from data police, including accident type, location, and number of vehicles involved. The traffic data collected include average annual daily traffic (AADT) collected from authorities like the Highway Planning Unit (HPU). Road features include roads and road characteristics such as length of tangent, length of curve, shoulder width, lane width, and access point. The data were collected from authorities such as the Public Ministry Work (JKR) or from site visits.

3 Result and Analysis

3.1 Development Operating Speed Model Tangent and Curve (85th Percentile)

The models were developed based on the new method of speed profile analysis. The profile was plotted based on continuous speed data obtained through GPS DG-200. The 85th percentile values were obtained from real speed and road elements through the analysis. By using statistical software (Minitab), the linear regression method was used to develop a model for prediction of the 85th percentile speed for the tangent and curve, in order to develop a model for the prediction of operating speeds of motorcycles and cars. Equations were developed using different variables such as length of tangent, length of curve, shoulder width, lane width, and access point. The statistical results are as follows;

Motorcycles

$$V_{85t}. = 72.0 + 11.7\,L_t. + 6.99\,SW - 2.61\,LW + 0.276\,AP \quad R^2 = 1 \tag{6}$$

$$V_{85c}. = 20.6 + 102\,L_c. + 4.16\,SW + 0.82\,AP \quad R^2 = 0.79 \tag{7}$$

Cars

$$V_{85t}. = 88.9 - 1.26\,L_t. + 27.4\,SW - 16.9\,LW + 1.07\,AP \quad R^2 = 0.99 \tag{8}$$

$$V_{85c}. = -31.1 + 202\,L_c + 4.1\,SW + 1.38\,AP \quad R^2 = 0.83 \tag{9}$$

Where;V_{85t} = Operating speed of tangent, V_{85c}. = operating eed of curve, L_t = length of tangent, L_c = length of curve, SW = shoulder width, LW = lane width, AP = access point.

3.2 Validation of Models

The validation of the developed models is carried out to verify t predictability of the model. Hence, scatter plot were plotted closed to the 4line between the observed speed and predicted operating speed model. The best fit linear relation between observed speed and predicted speed are shown in graph. Figure 2 shown comparisons observed speed vs. predicted speed of motorcycles a km21 until km25 while Fig. 3 shown comparisons observed speed vs. predicted speed of cars at km21 until km25.

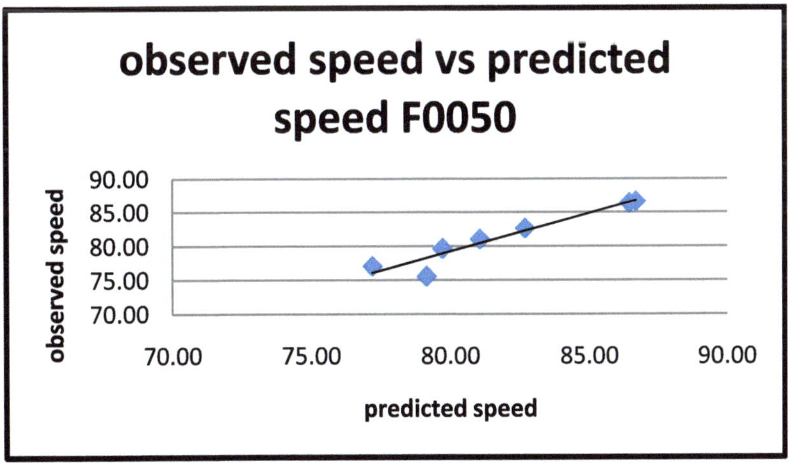

Fig. 2. Comparison observed speed vs. predicted speed of motorcycles (KM21–KM25)

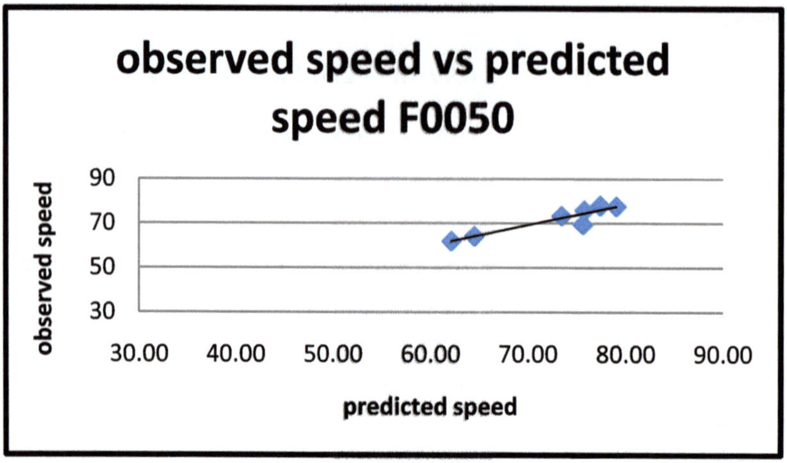

Fig. 3. Comparison observed speed vs. predicted speed of cars (KM21–KM25)

3.3 Development Design Consistency

The design consistency was determined based on continuous operating speed profile models. To develop the road design consistency model, ACAD software was used. Plots of the operating speed model profile (85th percentile) along the length are called speed profiles on this study. Figure 2 shows the speed profile F0050 of motorcycles at km21–km25. Figure 3 shows a speed profile F0050 of cars at km21–km25. The continuous operating speed profile models were plotted based on operating speed models (85th percentile).

Based on a profile that has been developed, two main parameters for this model can be determined. The bounded area between the profile, the average speed, and the standard deviation of speeds along a two-lane highway segment, was calculated to determine design consistency for every segment per km. Based on the two independent measures, a consistency model was developed. Table 1 shows design consistency motorcycles at km21–km25 per km. Table 2 shows design consistency cars at km21–km25 per km.

Table 1. Design consistency motorcycles (KM21–KM25) per km

Km	Design consistency	Threshold
21	1.02	Acceptable
22	1.55	Acceptable
23	0.55	Poor
24	0.53	Poor
25	2.10	Good

Table 2. Design consistency cars (KM21–KM25) per km

Km	Design consistency	Threshold
21	2.50	Good
22	1.64	Acceptable
23	0.55	Poor
24	0.33	Poor
25	0.46	Poor

3.4 Integrated Design Consistency Model

An integrated-consistency model is needed as a function of speed profiles for cars and motorcycles. The model is included with the impact of speed profiles for motorcycles on design consistency. Figure 4 and 5 show a speed profile F0050 of motorcycles and cars at km21–km25. Figure 6 show typical speed profiles of cars and motorcycles which created the normalized bounded area between the speed profiles of motorcycles and cars. Table 3 shows integrated design consistency between motorcycles and cars at km21–km25 per km. Every km shows the value of the integrated design consistency and the threshold at every km.

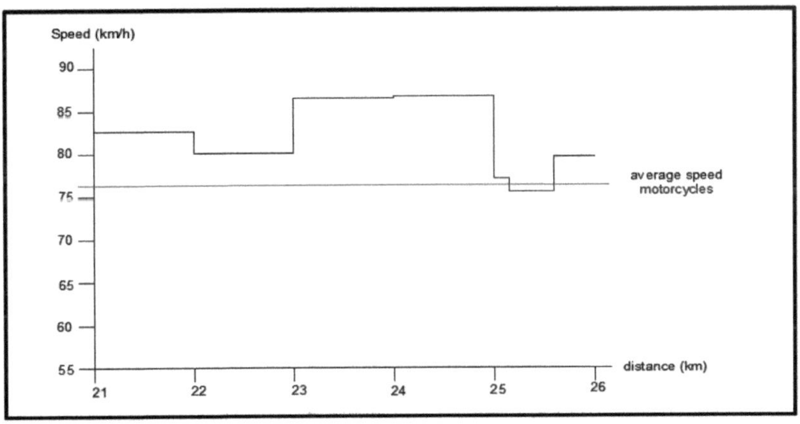

Fig. 4. Speed profile F0050 of motorcycles (KM21–KM25)

This study developed a new model to estimate the consistency of the segment at F0050. Based on two independents, the bounded area between the profile and the average speed and standard deviation of speeds, different approaches were given to develop a model. Hence the consistency is calculated for the selected segment, which is at km21 until km25.

The design consistency model was calculated at every km (km21 until km25). At km21, the threshold design consistency of motorcycles shows an acceptable design, while the design consistency of cars shows a good design. For the integrated design consistency motorcycle and cars, the threshold shows a good design for both types of

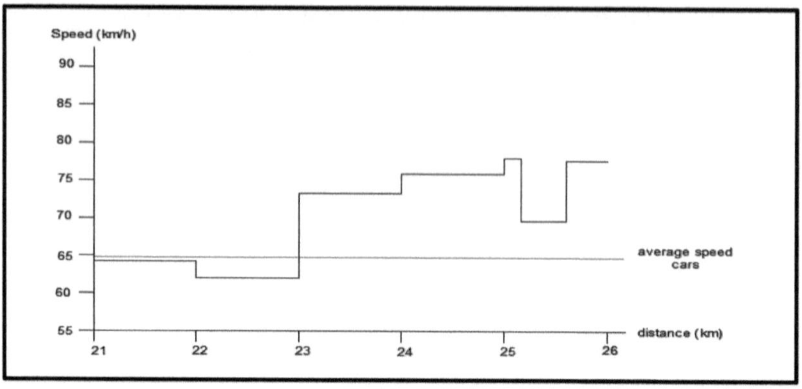

Fig. 5. Speed profile F0050 of cars (KM21–KM25)

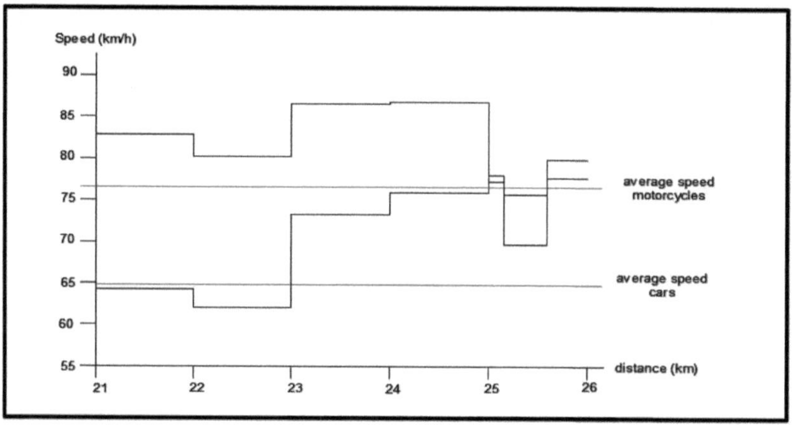

Fig. 6. Speed profile F0050 of motorcycles and cars (KM21–KM25)

vehicles. At km22, the threshold design consistency of motorcycles and design consistency of cars show as acceptable. For the integrated design consistency of motorcycle and cars, the threshold shows an acceptable design for both types of vehicles. At km23 and km24, the threshold design consistency of motorcycles and the design consistency of cars shows a poor design. For the integrated design consistency of motorcycle and cars, the threshold shows a poor design for both types of vehicles. At km25, the threshold design consistency of motorcycles shows a good design, while the design consistency of cars shows a poor design; however, for the integrated design consistency of motorcycles and cars, the threshold shows a poor design for both types of vehicles.

In police data according to statistics over 5 years, km24 has the highest area of accidents that can be explained by the design consistency of the area based on an integrated design consistency model between cars and motorcycles which is of poor

Table 3. Integrated design consistency motorcycle and cars (KM21 – KM25)

Km	Integrated design consistency	Threshold	APW
21	2.08	Good	26.0
22	1.37	Acceptable	11.8
23	0.48	Poor	12.8
24	0.30	Poor	72.4
25	0.45	Poor	51.8

design. Km24 was the busiest area with factors which can be related to the many access points, its location around the shop, and its lack of a median in the middle of the road along segment km24. These factors can be conclusive causes of the accidents which are highest in the area km24.

Based on APW, police data show km23 experienced a low accident area, but the design consistency of the area based on the integrated design consistency model between cars and motorcycles is of poor design. It can be concluded that speed is not the fault of accidents which occur, but rather the factors of the surrounding area and driver behavior itself.

4 Conclusion

Continuous speed profile data from GPS are being used to develop design consistency profiles for motorcycles and cars by developing the 85th percentile operating speed model for motorcycles and cars. The 85th percentile operating speed model was developed for tangent and curve using the parameter road features. The design consistency model is based on two independents, the bounded area between the profiles and the average speed and standard deviation of speeds.

Based on the results and analysis, the conclusion of the result is that if the integrated design consistency between motorcycles and cars shows a poor design consistency, the potential for accidents is higher. Even with good design consistency, accidents can occur. Design consistency influences drivers to drive with safe conditions. Further studies can estimate the occurrence of motorcycle and car accidents using Poisson distribution along the selected segment at F0050.

Acknowledgments. The authors would like to thank the Ministry of Higher Education of Malaysia for supporting the study through FRGS grant Vot. 1452 (Ref: FRGS/1/2014/TK07/UTHM/02/2) and all individuals/organization that have made this study possible. Thanks are also extended to the authorities of Faculty of Civil Engineering and Environment, Universiti Tun Hussein Onn Malaysia.

References

Abdullah, L., Zamri, N.: Road accident models with two threshold levels of fuzzy linearregression. J. Emerg. Trends Comput. Inform. Sci. **3**(2), 225–230 (2012)

Castro, M., Iglesias, L., Rodriguez-Solano, R., Sanchez, J. A.: Highway safety analysis using geographic information systems. In: Proceedings of the Institution of Civil Engineers-Transport, vol. 161, no. 2, pp. 91–97. Thomas Telford Ltd., May 2008

De Oña, J., Garach, L.: Accidents Prediction Model based on Speed Reduction on Spanish Two-Lane Rural Highways. Proc.-Soc. Behav. Sci. **53**, 1010–1018 (2012)

Gibreel, G.M., Easa, S.M., Hassan, Y., El-Dimeery, I.A.: State of the art of highway geometric design consistency. J. Transp. Eng. **125**(4), 305–313 (1999)

Hung, D., Stevenson, M., Ivers, R.: Prevelance of helmet use among motorcycle riders in Vietnam. Inj. Prev. **12**(6), 409–413 (2006)

Lamm, R., Psarianos, B., Mailaender, T.: Highway Design and Traffic Safety Engineering Handbook (1999)

Mattar-Habib, C., Polus, A., Farah, H.: Further evaluation of the relationship between enhanced consistency model and safety of two-lane rural roads in Israel and Germany. EJTIR, **4**(8) (2008)

Memon, R.A., Khaskheli, G.B., Dahani, M.A.: Estimation of operating speed on two lane two way roads along N-65 (SIBI–Quetta). Int. J. Civ. Eng. **10**(1), 25–31 (2012)

Polus, A., Mattar-Habib, C.: New consistency model for rural highways and its relationship to safety. J. Transp. Eng. **130**(3), 286–293 (2004)

Prasetijo, J., Zainal, Z.F.: Development of continuous speed profile using GPS at Johor Federal Roads F0050. In: MATEC Web of Conferences, vol. 47. EDP Sciences, January 2016

Prasetijo, J., Zainal, Z.F., Musa, W.Z.: Speed profile based on design consistency. In: Proceedings International Conference on Engineering of Tarumanagara, vol. 1, no. 1 (2015)

Radin, U.: Updates of road safety status in Malaysia. IATSS Res. **29**, 78–80 (2005)

Influence of Asymmetrical Topology on Structural Behaviours of Bearers and Sleepers in Turnout Switches and Crossings

Sakdirat Kaewunruen[1], Alex M. Remennikov[2], and Serdar Dindar[3(✉)]

[1] Railway and Civil Engineering, School of Engineering, The University of Birmingham, Birmingham B152TT, UK
s.kaewunruen@bham.ac.uk
[2] School of Civil Mining and Environmental Engineering, University of Wollongong, Wollongong, NSW 2502, Australia
[3] Birmingham Centre for Railway Research and Education, School of Engineering, The University of Birmingham, Birmingham B152TT, UK
SXD319@bham.ac.uk

Abstract. Railway infrastructure is nonlinear by nature, scientifically proven by its behaviours, geometry and alignment, wheel-rail forces and operational parameters such as tractive efforts. It is often found that most train-turnout interaction models do not consider the time dependent ballast degradation. Such ballast degradation later causes differential settlement and aggravates impact forces acting on partial and unsupported sleepers and bearers. Furthermore, localised ballast breakages underneath any railseat increase the likelihood of centre-bound cracks in railway sleepers and bearers due to the unbalanced support. This paper presents a numerical simulation of a standard-gauge concrete bearer at crossing panel, taking into account the tensionless nature of ballast support. The finite element model was calibrated using static and dynamic responses using past experiments. In this paper, the influences of topologic asymmetry on both sagging and hogging behaviours of crossing bearers are firstly investigated. In addition, it is the first to demonstrate the effects of sleeper length on the design consideration of turnout bearers in crossing panel. The outcome of this study will improve the railway turnout construction and maintenance criteria in order to improve train-turnout interaction and ride comfort.

1 Introduction

In ballasted railway tracks, railway sleepers (also called 'railroad tie' in North America) are a vital element of railway track structures. Their key role is to redistribute loads from the rails to the underlying ballast bed. Based on the current design approach, the design life span of the concrete sleepers is targeted at around 50 years in Australia and around 70 years in Europe (Australian Standards 2003; Kaewunruen et al. 2015, 2016a,

J. Pombo and G. Jing (eds.), *Recent Developments in Railway Track and Transportation Engineering*, Sustainable Civil Infrastructures, DOI 10.1007/978-3-319-61627-8_4

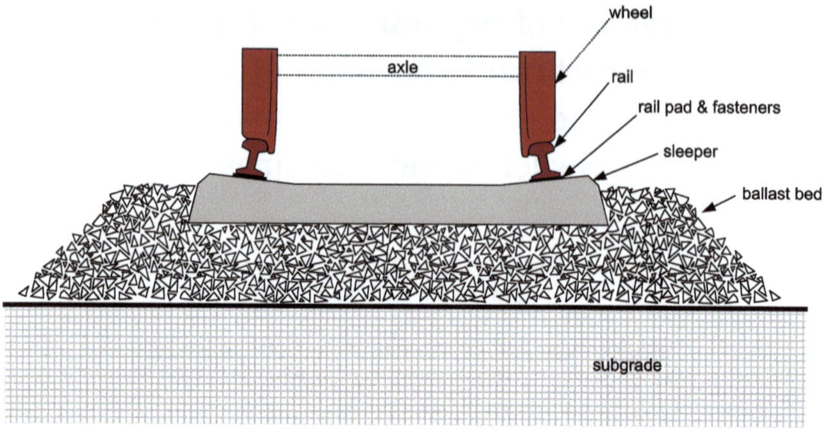

Fig. 1. Typical ballasted railway track components.

2016b). Figure 1 shows the typical ballasted railway tracks and their key components. There have been a number of previous investigations on the railway sleeper models (Neilsen 1991; Cai 1992; Grassie 1995). Most of the models employed the concept of beam on elastic foundation where a sleeper is laid on the elastic support, acting like a series of springs. It is found that only vertical stiffness is sufficient to simulate the ballast support condition because the lateral stiffness seems to play an insignificant role in sleeper's bending responses (Cai 1992; Kaewunruen and Remennikov 2006, 2007). In practice, the lateral force is less than 20% of vertical force and the anchorage of fastening has been designed to take care of lateral actions (Remennikov and Kaewunruen 2008). In fact, field measurements suggest a diverse range of sleeper flexural behaviors, which are largely dependent on the support condition induced by ballast packing and tamping (Wolf et al. 2015; Sae Siew et al. 2015; Vu et al. 2016). However, it is still questionable at large whether modern ballast tamping process is effective and it could enable adequate symmetrical support for sleeper at railseat areas. In reality, the ballast is tamped only at the railseat areas. The ballast at the mid span is left loosening, with the intention to reduce negative bending moment effect on sleeper mid span, which is the cause of centre bound. Over time, the dynamic track settlement induces ballast densification and the sleeper mid-span comes into contact or is fully supported by ballast until the track geometry is restored by resurfacing activity (i.e. re-tamping).

In contrast, the structural behavior of turnout bearers has not been fully investigated. Figure 2 shows the typical layout of a turnout system. A railway turnout system have generally been analysed the using a grillage beam method (Ferdous et al. 2015). Although the simplification is useful, such a method could not adequately assist in the failure analyses of turnout components. In some cases, the results using the grillage beam method seem to have discrepancies with the field observations where the maximum bending and shear forces were evident within the crossing panel (Kaewunruen 2012). A number of research has been conducted to locate the critical section within a

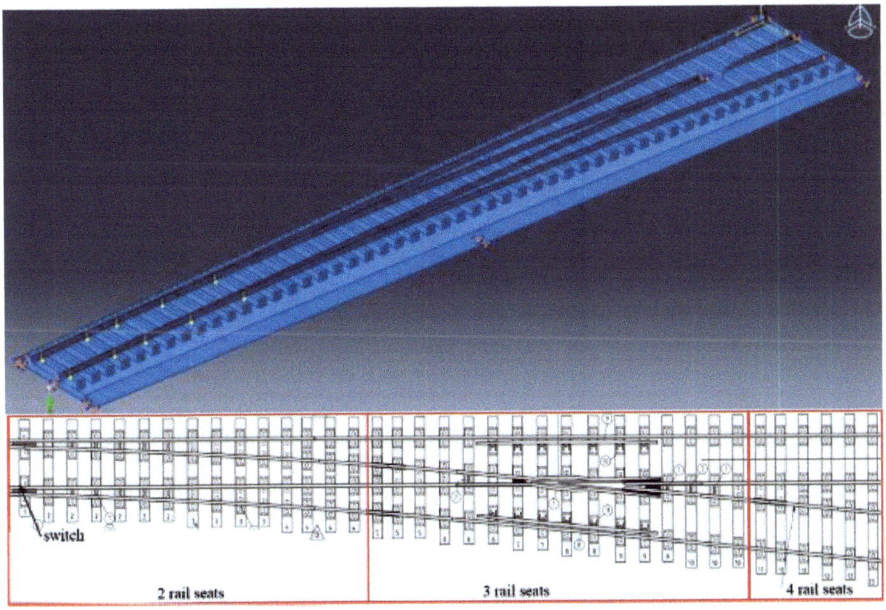

Fig. 2. Typical turnout system layout (Sae Siew et al. 2016).

turnout, and many of which conclude that the critical section is located specifically at the crossing panel at either v-crossing or k-crossing (Kassa and Nielsen 2009; Wiest et al. 2008; Xiao et al. 2011).

Although it is clear that the turnout bearers are topological asymmetry, such the aspect has never been fully investigated. This paper presents an advanced railway concrete sleeper modeling capable of analysis into the effect of topological asymmetry on the positive and negative flexural responses of railway sleepers. It focuses on the nonlinear static flexural response of railway concrete sleepers subjected to a spectrum of ballast stiffness at the mid span, in comparison with the current design method in accordance with the design standards.

2 Finite Element Model

Previous extensive studies established that the two-dimensional Timoshenko beam model is the most suitable option for modeling concrete sleepers under vertical loads (Nielsen 1991; Grassie 1995). In this investigation, the finite element model of concrete sleeper (optimal length) has been previously developed and calibrated against the numerical and experimental modal parameters (Kaewunruen and Remennikov 2006a). Figure 3 shows the two-dimensional finite element model for an in-situ railway concrete sleeper. Using a general-purpose finite element package STRAND7 (G+D Computing 2001), the numerical model included the beam elements, which take into account shear and flexural deformations, for modeling the concrete sleeper. The trapezoidal

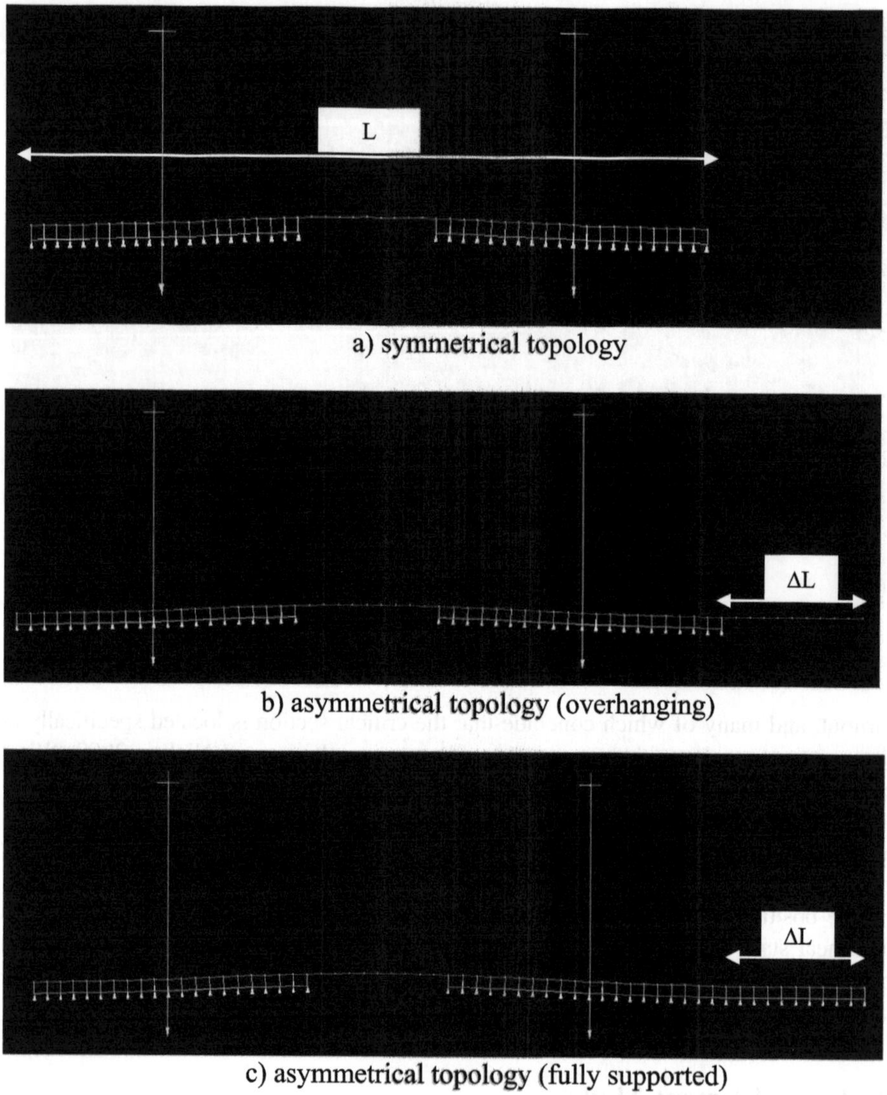

Fig. 3. STRAND7 finite element model of a concrete bearer

cross-section was assigned to the sleeper elements. The rails and rail pads at railseats were simulated using a series of spring. In this study, the sleeper behaviour is stressed so that very small stiffness values were assigned to these springs. In reality, the ballast support is made of loose, coarse, granular materials with high internal friction. It is often a mix of crushed stone, gravel, and crushed gravel through a specific particle size distribution. It should be noted that the ballast provides resistance to compression only.

As a result, the use of elastic foundation in the current standards in Australia and North America (Standards Australia 2003; Wolf et al. 2015) does not well represent the real uplift behaviour of sleepers in hogging moment region (or mid span zone of railway sleeper). In this study, the support condition was simulated using the tensionless beam support feature in Strand7 (G+D Computing 2001). This attribute allows the beam to lift or hover over the support while the tensile supporting stiffness is omitted. The tensionless support option can correctly represent the ballast characteristics in real tracks (G+D Computing 2001). Table 1 shows the geometrical and material properties of the finite element model. It is important to note that the parameters in Table 1 give a representation of a specific rail track. These data have been validated and the verification results have been presented elsewhere (Kaewunruen et al. 2015).

Table 1. Engineering properties of the standard sleeper used in the modeling validation

Parameter lists		
Flexural rigidity	$EI_c = 4.60$, $EI_r = 6.41$	MN/m^2
Shear rigidity	$\kappa GA_c = 502$, $\kappa GA_r = 628$	MN
Ballast stiffness	$k_b = 13$	MN/m^2
Rail pad stiffness	$k_p = 17$	MN/m
Sleeper density	$\rho_s = 2{,}750$	kg/m^3
Sleeper length	$L = 2.5$	m
Rail gauge	$g = 1.5$	m

Based on our critical literature review, the flexural influences on railway concrete bearers in a turnout system (switch and crossing) due to the variations of ballast support conditions together with the asymmetric topology of sleeper has not yet addressed by the past researchers (Manalo et al. 2012). Especially when the uplift behaviour due to ballast tensionless support in hogging region of sleepers is considered, a finite element analysis is thus required to supersede the simple manual calculation. For this study, the numerical simulations have been extended to conduct the analyses using the nonlinear solver in STRAND7. The effects of asymmetric topology of concrete bearers on their flexural responses in a turnout system can be evaluated. The length of bearer varies from 2.5 m to 4.0 m, which is practically common in the 2 and 3 rail-seats sections (see Fig. 2).

3 Results and Discussion

Using the design data in Tables 1, 2 and 3 present the static bending moment envelops along the bearer when subjected to the equal wheel loads of 100 kN at both railseats, in comparison with the standard design moments. Based on AS1085.14 (Standards Australia 2003), the design maximum positive bending moment at the rail seat = 12.50 kNm, while the centre negative design bending moment = 6.95 kNm (if considered half support) or = 12.50 kNm (if considered full support). It is typical that the

Table 2. Maximum bending moment of overhanging bearer

ΔL/L (overhanging)	At railseat (kNm)		At mid span (kNm)	
	M*	M^*/M_{Design}	M*	M^*/M_{Design}
0	+11.93	0.95	−0.95	0.14
10%	+11.93	0.95	−0.95	0.14
20%	+11.93	0.95	−0.96	0.14
30%	+11.93	0.95	−0.96	0.14
40%	+11.93	0.95	−0.96	0.14
50%	+11.93	0.95	−0.96	0.14
60%	+11.93	0.95	−0.96	0.14

Table 3. Maximum bending moment of fully-supported bearer

ΔL/L (full support)	At railseat (kNm)		At mid span (kNm)	
	M*	M^*/M_{Design}	M*	M^*/M_{Design}
0	+11.93	0.95	−0.95	0.14
10%	+15.16	1.21	+2.22	0.32
20%	+16.50	1.32	+3.15	0.45
30%	+16.74	1.34	+3.29	0.47
40%	+16.74	1.34	+3.29	0.47
50%	+16.74	1.34	+3.29	0.47
60%	+16.74	1.34	+3.30	0.47

positive and negative moments are associated with the railseat and mid-span sections, respectively. It shows that the standard design moments provide the conservative results. The standard design moment at mid span is about half between the other two cases (see Fig. 3).

Based on the static results in Tables 2 and 3, it is clear that the influence of the asymmetrical topology is pronounced when there is a contact between bearer and ballast layer. Considering field investigation, such the contact could occur when there is a differential settlement on the mainline track (or run-through turnout road). Once the ballast-bearer contact establishes, the bearer will take additional bending moment at the inner railseat.

The natural frequencies of the bearers can be observed in Tables 4 and 5. It can be seen that the topology of bearer plays a key role in dynamic natural frequencies and corresponding mode shapes of the bearers. Overhanging bearers tend to be relatively much affected by the topology aspect in comparison with the dynamic behavior of fully supported bearers. Figure 4 shows the dynamic softening behavior of the turnout bearers with asymmetrical topology. It is clear that the dynamic softening is more pronounced at a higher frequency range.

Table 4. Natural frequencies of overhanging bearer

ΔL/L (overhanging)	Resonances (Hz)			
	Mode 1	Mode 2	Mode 3	Mode 4
0	143 	370 	714 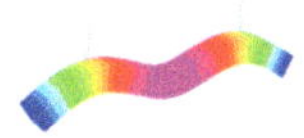	1155 
10%	118 	308 	594 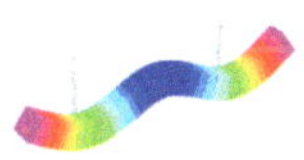	957 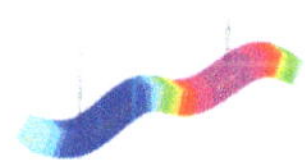
20%	102 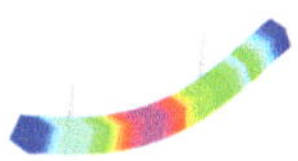	261 	498 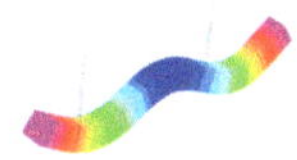	818 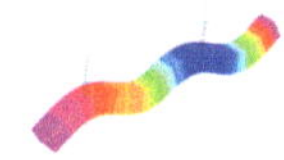
30%	89 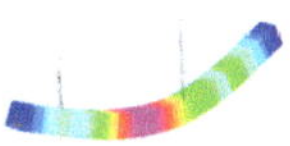	224 	426 	702 
40%	80 	194 	369 	601 
50%	73 	170 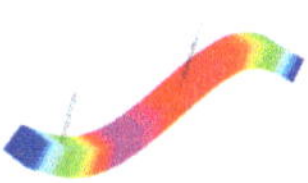	322 	524 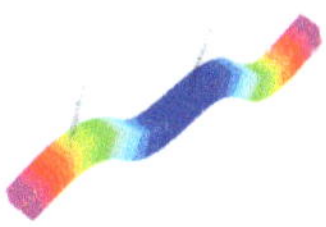
60%	67 	150 	284 	470 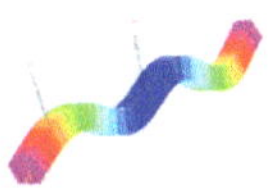

Table 5. Natural frequencies of fully-supported bearer

ΔL/L (full support)	Resonances (Hz)			
	Mode 1	Mode 2	Mode 3	Mode 4
0	143	370	714	1155
10%	121	309	599	970
20%	105	263	504	830
30%	93	227	431	716
40%	84	197	374	620
50%	77	173	328	542
60%	71	154	290	480

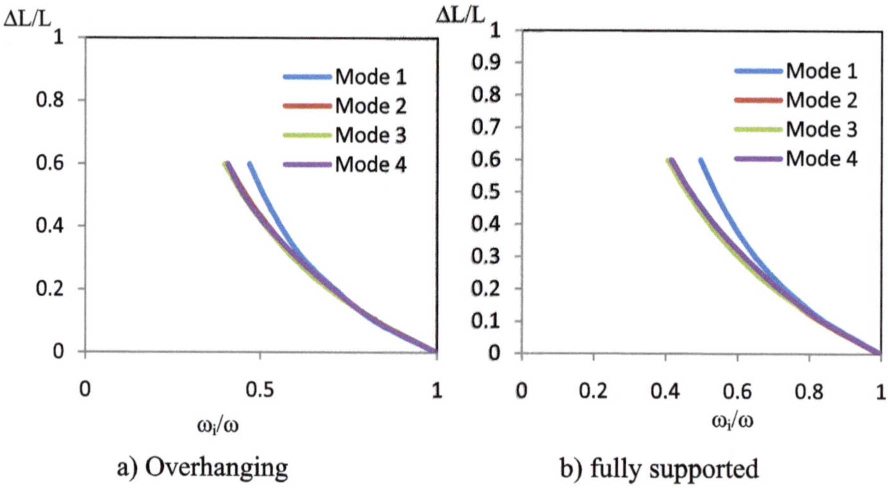

Fig. 4. Frequency ratios of turnout bearer

4 Conclusions

This paper numerically investigates the critical structural effects of a variety of ballast conditions and asymmetric topology on the flexural responses and free vibration behaviors of the railway sleepers and bearers in a turnout system (switches and crossings). The finite element model of bearers, which was established and calibrated earlier, is utilised in this study. The influences of the variation of ballast support conditions at bearer end together with the asymmetric length of sleepers on the bending of the railway sleeper were highlighted in comparison with the standard design. The nonlinear solver in STRAND7 was employed to handle sleeper/ballast contact mechanics. Under static and free vibration conditions for overhanging and supported bearers, the numerical results exhibit that the bending moment resultants are barely affected by topological aspects when the ballast-sleeper contact is not established. The standard design bending moments tend to be overestimated for the overhanging bearer, whilst they can be highly underestimated when bearer end is laid on ballast. Generally, positive bending moments at inner railseat of bearer have generally high sensitivity to the spectrum of ballast support conditions in comparison with the more pronounced influence of sleeper length. In such case, the nominal bending moment at inner railseat could be larger than the structural capacity of sleeper and resulted in structural cracks and failure. In contrast, such behavior is insignificant and tolerable for overhanging bearers. By understanding the free vibration behavior of bearers, it is clear that the asymmetrical topology induces dynamic softening in the turnout bearers. This implies that the asymmetrical bearers are prone to damage under high-intensity impact loading, which could trigger and sweep through various resonant frequencies of the turnout bearers. The insight in this structural behavior of bearer has raised the awareness of track engineers for better design and maintenance of switch and crossing support structures.

Acknowledgments. The first author wishes to gratefully acknowledge the Japan Society for Promotion of Science (JSPS) for his JSPS Invitation Research Fellowship (Long-term), Grant No L15701, at Track Dynamics Laboratory, Railway Technical Research Institute and at Concrete Laboratory, the University of Tokyo, Tokyo, Japan. The JSPS financially supports this work as part of the research project, entitled “Smart and reliable railway infrastructure”. Special thanks to European Commission for H2020-MSCA-RISE Project No. 691135 “RISEN: Rail Infrastructure Systems Engineering Network”. In addition, the sponsorships and assistance from CEMEX, Network Rail, RSSB (Rail Safety and Standard Board, UK) and G+D Computing (Dr Erik Kostson) with respect to STRAND7 are highly appreciated.

References

Standards Australia: Australian Standard: AS1085.14-2003 Railway track material - Part 14: Prestressed concrete sleepers, Sydney, Australia (2003)

Neilsen, J.C.O.: Eigenfrequencies and eigenmodes of beam structures on an elastic foundation. J. Sound Vibr. **145**, 479–487 (1991)

Cai, Z.: Modelling of rail track dynamics and wheel/rail interaction, Ph.D. Thesis, Department of Civil Engineering, Queen’s University, Ontario, Canada (1992)

G+D Computing: Using Strand7: Introduction to the Strand7 finite element analysis system, Sydney, Australia (2001)

Grassie, S.L.: Dynamic modelling of concrete railway sleepers. J. Sound Vibr. **187**, 799–813 (1995)

Kassa, E., Nielsen, J.C.O.: Dynamic train–turnout interaction in an extended frequency range using a detailed model of track dynamics. J. Sound Vib. **320**, 893–914 (2009)

Kaewunruen, S.: Effectiveness of using elastomeric pads to mitigate impact vibration at an urban turnout crossing. In: Maeda, T., et al. (eds.) Noise and Vibration Mitigation for Rail Transportation Systems, pp. 357–365. Springer, Tokyo (2012)

Kaewunruen, S., Remennikov, A.M.: Sensitivity analysis of free vibration characteristics of an in-situ railway concrete sleeper to variations of rail pad parameters. J. Sound Vibr. **298**(1), 453–461 (2006a)

Kaewunruen, S., Remennikov, A.M.: Rotational capacity of railway prestressed concrete sleeper under static hogging moment. In: the 10th East Asia-Pacific Conference on Structural Engineering and Construction, Bangkok, Thailand (2006b)

Kaewunruen, S., Remennikov, A.M.: Investigation of free vibrations of voided concrete sleepers in railway track system. Proc. Inst. Mech. Eng. Part F J. Rail Rapid Transit **221**(4), 495–507 (2007)

Kaewunruen, S., Ishida, T., Remennikov, A.M.: Numerical simulations of negative flexural responses (hogging) in railway prestressed concrete sleepers. In: RILEM International Conference on Strategies for Sustainable Concrete, Rio de Janeiro, Brazil, 14–16 December 2015

Kaewunruen, S., Ishida, T., Remennikov, A.M.: Impact analyses for negative flexural responses (hogging) in railway prestressed concrete sleepers. In: International Conference on Recent Advances in Structural Dynamics, Southampton, UK, 3–6 July 2016 (2016a)

Kaewunruen, S., Minoura, S., Watanabe, T., Remennikov, A.M.: Remaining service life of railway prestressed concrete sleepers. In: Proceedings of International RILEM Conference on Materials, Systems and Structures in Civil Engineering, 22–24 August 2016, Technical University of Denmark, Lyngby, Denmark (2016b)

Kaewunruen, S., Chamniprasart, K.: Damage analysis of spot replacement sleepers interspersed in ballasted railway tracks. In: Proceedings of the 29th Nordic Conference on Computational Mechanics, Chalmers University of Technology, Gotenburg, Sweden (2016)

Manalo, A., Aravinthan, T., Karunasena, W., Stevens, N.: Analysis of a typical railway turnout sleeper system using grillage beam analogy. Finite Elem. Anal. Des. **48**(1), 1376–1391 (2012)

Remennikov, A.M., Kaewunruen, S.: A review on loading conditions for railway track structures due to wheel and rail vertical interactions. Struct. Control Health Monit. **15**(2), 207–234 (2008)

Sae Siew, J., Mirza, O., Kaewunruen, S.: Nonlinear finite element modelling of railway turnout system considering bearer/sleeper-ballast interaction. J. Struct. (2015). http://dx.doi.org/10.1155/2015/598562

Sae Siew, J., Mirza, O., Kaewunruen, S.: Torsional effect on track support structures of rail turnouts crossing impact. ASCE J. Transp. Eng. (2016, in press)

Vu, M., Kaewunruen, S., Attard, M.: Chap. 6 – Nonlinear 3D finite-element modeling for structural failure analysis of concrete sleepers/bearers at an urban turnout diamond. In: Handbook of Materials Failure Analysis with Case Studies from the Chemicals, Concrete and Power Industries, pp. 123–160. Elsevier, The Netherlands (2016). http://dx.doi.org/10.1016/B978-0-08-100116-5.00006-5

Wiest, M., Kassa, E., Daves, W., Nielsen, J.C.O., Ossberger, H.: Assessment of methods for calculating contact pressure in wheel-rail/switch contact. Wear **265**, 1439–1445 (2008)

Wolf, H.E., Edwards, J.R., Dersch, M.S., Barkan, C.P.L.: Flexural analysis of prestressed concrete monoblock sleepers for heavy-haul applications: methodologies and sensitivity to support conditions. In: Proceedings of the 11th International Heavy Haul Association Conference, Perth, Australia (2015)

Xiao, J., Zhang, F., Qian, L.: Numerical simulation of stress and deformation in a railway crossing. Eng. Fail. Anal. **18**, 2296–2304 (2011)

Geological Hazard Risk Evaluation for Railway Network of Guizhou Province in China

Rui Tang[1(✉)], Weidong Wang[1,2], Jie Ma[3], and Yanping Chen[4]

[1] School of Civil Engineering, Central South University, Changsha, Hunan 410075, China
147745@163.com
[2] National Engineering Laboratory for High Speed Railway Construction, Changsha, Hunan 410075, China
[3] China Railway Eryuan Engineering Group Company Limited, Chengdu, Sichuan 610031, China
[4] China Railway Siyuan Survey and Design Group Company Limited, Wuhan, Hubei 430063, China

Abstract. In recent years, China's high-speed railway has experienced a period of rapid development and being gradually rational. This paper took Guizhou Province as the study area, one of the places in China which are most seriously affected by landslide hazards. The research in this paper was conducted in three steps. Firstly, the landslide susceptibility mapping of railway was acquired by applying competition network model, and a set of conditioning factors were selected as the major landslide-conditioning factors, including elevation, lithology, rainfall, distance from river, distance from tectonic line, karst density and slope. Then, the concept of 'degree of fitting' was proposed in the assessment of railway risk degree, and it was regarded as one of the three elements which determine the railway protection grade on geological disasters. Finally, the matter-element model was established based on extension method, which can be used to evaluate the protection grades for the planned railway on geological disasters by integrating three elements, the train speed, grade of susceptibility mapping, and fitting degree, into the model.

1 Introduction

The research on geological disaster susceptibility mapping began in the 1960s. A considerable amount of research works have been conducted over the past years. Current method for geological disaster susceptibility mapping can be categorized into two groups, i.e., qualitative and quantitative. Qualitative methods are mainly on the basis of some fuzzy approaches, combining with the experiences of the experts to make decisions, and quantitative methods generally rely on mathematical models or some software or data equations to calculate. In abroad, Varnes studied the type and movement processes of slope geological disasters, and put forward early analysis and control methods [1]; Pistocchi et al. established a landslide susceptibility map with expert system method, achieving good results [2]; The Turkish scholar Yesilnacar

J. Pombo and G. Jing (eds.), *Recent Developments in Railway Track and Transportation Engineering*, Sustainable Civil Infrastructures, DOI 10.1007/978-3-319-61627-8_5

utilized artificial neural network model and set up a landslide hazard susceptibility map for Turkish [3]; Melchiorre et.al studied the landslide sensitivity evaluation on the basis of neural network model and cluster analysis [4]. In our country, many research methods about geological hazard susceptibility mapping have also been carried out in recent years. Yin Kunlong at China University of Geosciences studied the alarm and forecast system of sudden geological disasters in Zhejiang Province [5]; Dai Fuchu summarized the various research methods on landslide disasters [6]. Most articles are about susceptibility mapping. The innovation of this article is putting forward of 'fitting degree' concept and establishing the quantitative method of matter-element for evaluating railway protection grade. The flowchart of this paper is shown in Fig. 1.

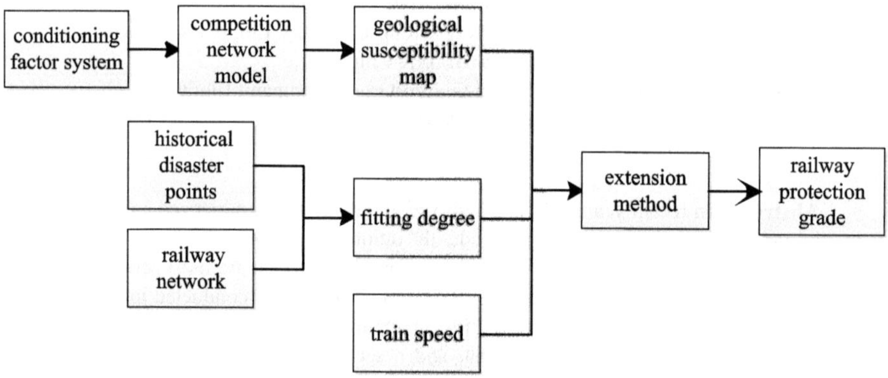

Fig. 1. The flowchart of geological hazard risk evaluation for railway network

2 Geological Hazard Susceptibility Evaluation

2.1 The Selection of Landslide-Causing Factors

According to the geology, topography and other natural conditions, this paper comprehensive selected seven conditioning factors including elevation, lithology, rainfall, distance from river, distance from tectonic line, karst density and slope to determine the geological hazard susceptibility mapping. In order to classify and quantify these hazard factors, this research divided them into several sub-classes. The area of history disaster within each sub-class was calculated and normalized.

2.2 The Determination of Factor Weight Based on Competition Network

Competition network consists of a set of layers, namely, input layer, hidden layer, and output layer (Fig. 2). Each layer in the competition network consists of independent processing units called neurons. These neurons are linked to neurons in other layers through the weights and bias.

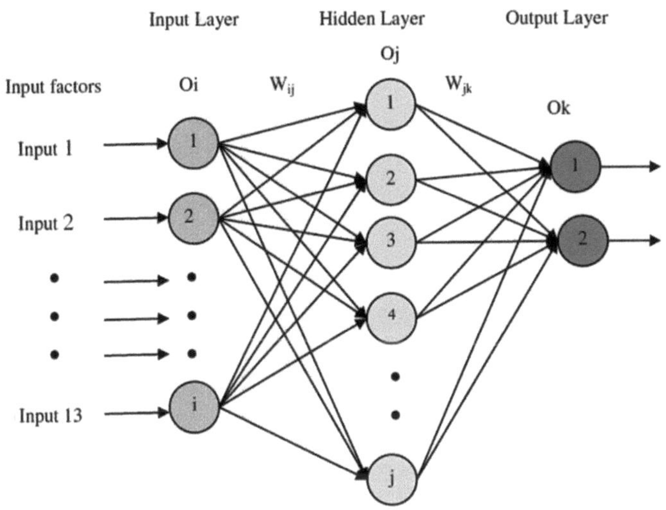

Fig. 2. Architecture of competition network

The normalized data of the seven factors were regarded as the input layers, and the landslide risk degrees as the output layers.

By staking the seven factors together in AcrMap platform, the original layer of each factor was divided into finer and irregular polygons, which contained all normalized data of the seven factors. This paper took the whole Guizhou Province as the study area which was divided into 4073 small polygons, from which twelve typical polygons were selected as the test data. After repeatedly adjusting the weight parameters, the factor weights were acquired through adjusting the 12 polygons to be classified accurately. By using these well-trained data, the whole dataset of Guizhou Province was imported after being trained in one time based on the Matlab software.

2.3 Geological Susceptibility Map of Guizhou Province

After importing the Matlab calculating results into the DBF tables in ArcGIS and rendering different sensitivity degrees based on ArcMap, the landslide susceptibility map is obtained (Fig. 3). According to the difference in landslide susceptibility degree, the study area was divided into five grades: the very low susceptible zone VLS), the low susceptible zone (LS), the moderate susceptible zone (MS), the moderate-high susceptible zone (MHS) and the high susceptible zone (HS). The statistics of the historical disaster area among each risk degree are shown in Table 1.

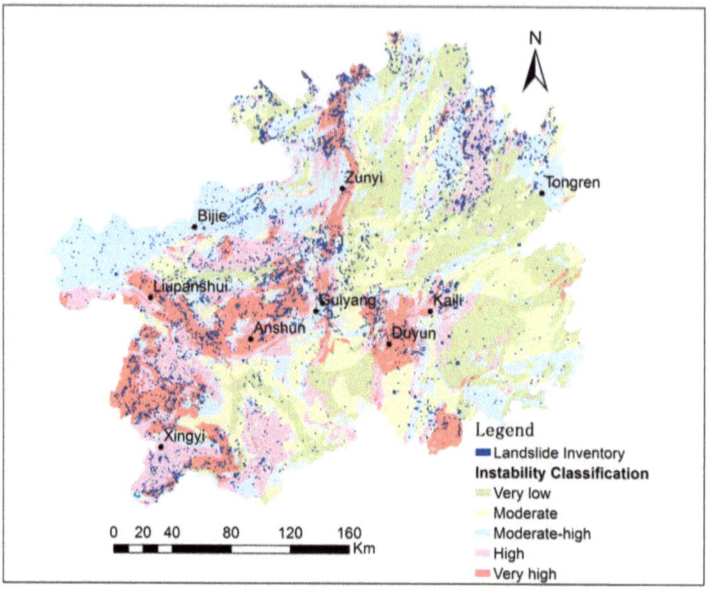

Fig. 3. Landslide susceptibility map of Guizhou Province

Table 1. Landslide distribution in landslide susceptible zones

Landslide susceptible zones	Area		
	Historical landslide area	Coverage area	Landslide density (%)
VLS	1077.63	38659.1	2.79%
LS	1636.20	36190.8	4.52%
MS	2634.69	42506.1	6.20%
MHS	3266.86	33510.1	9.75%
HS	3626.65	25228.9	14.37%

3 The Geological Hazard Risk Assessment for Railway Network in Guizhou Province

According to the geological susceptibility map and historical landslide hazard distribution, we obtained the risk degrees along the planning railway line and got the distribution of landslide points on both sides of railways (Fig. 4). According to the distribution proportion among five landslide risk degrees, each railway was endowed a hazard weight. The five grade weight coefficient $\theta = (7, 5, 3, 2, 1)$. For example, the hazard weight of Hu-Kun passenger special line was:

$$W_1 = w\theta^T = (35.6\%, 25.4\%, 17.3\%, 18.9\%, 2.8\%) \cdot (7, 5, 3, 2, 1)^T = 4.69.$$

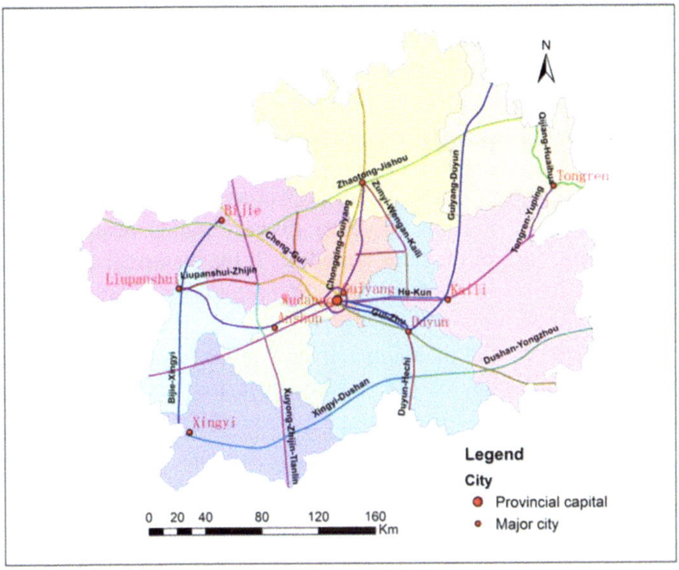

Fig. 4. Planning railways in Guizhou Province

Table 2. The distribution ratio and hazard weight of planning railway lines

Railway lines	Risk degree					
	HS (%)	MHS (%)	MS (%)	LS (%)	VLS (%)	Hazard weight
Hu-Kun	35.6	25.4	17.3	18.9	2.8	4.69
Cheng-Gui	25.1	33.5	32.1	2.5	6.8	4.51
Bijie-Xingyi	48.3	15.1	10.4	20.4	5.8	4.91
Liupanshui-Guiyang-Zunyi	50.0	33.1	11.0	5.7	0.2	5.60
Guiyang-Duyun	15.3	14.8	13.2	50.1	6.6	3.28
Guiyang-Kaili	30.5	20.8	24.5	16.8	7.4	4.32
Tongren-Yuping	0	0	6.8	10.3	82.9	1.24
Zhaotong-Jishou	8.7	10.9	61.5	10.1	8.8	3.29
Liupanshui-Jinzhi	17.6	15.7	41.3	20.4	5.0	3.71
Guiyang-Duyun-Enshi	9.8	18.9	24.0	10.2	37.1	2.93
Duyun-Hechi	33.5	10.2	5.3	51.0	0	4.03
Wengan-Fuquan	0	0	10.0	42.1	47.9	1.62
Xuyong-Zhijin-Tianlin	5.1	45.7	13.1	15.4	20.7	3.55
Jinsha-Xindian	40.6	49.1	1.3	0	9.0	5.43
Xingyi-Dushan	15.9	29.2	3.4	22.5	29.0	3.42
Chongqing-Guiyang	73.2	10.3	11.5	1.9	3.1	6.05
Qianjing-Huaihua	1.7	5.7	70.8	9.5	12.3	2.84
Dushan-Yongzhou	6.8	5.9	22.8	30.9	33.6	2.41
Zunyi-Wengan-Kaiyang	1.2	2.8	7.9	50.1	38.0	1.84
Huangtong-Zhijin	75.6	13.2	11.2	0	0	6.29
Gui-Zhu	19.9	6.5	12.4	25.7	35.5	2.96

The distribution of the 21 planning railway lines in Guizhou Province among 5 landslide risk degrees was shown in Table 2.

3.1 Assessment of Railway Risk Degree Based on Fitting Degree

By using the concept of 'fitting degree' [7–9], it was assumed that each railway line is a regression line (or curve), and the historical disaster points nearby the railway are discrete points on both sides of the regression line. If the historical disaster points more densely distributes in the vicinity of the line, the fitting degree will be greater, and the risk of the railway will be higher.

This paper extracted all the planning railways in Guizhou Province, and took the buffer of 10 km on both sides of the railway lines for research based on ArcMap software (Fig. 5).

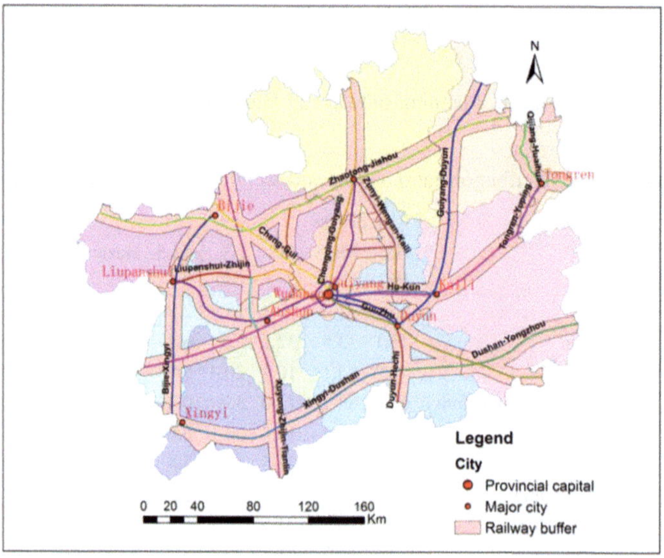

Fig. 5. The landslide hazard distribution within railway buffer of 10 km

The fitting degrees of each railway were calculated in the attribute Table 3.

Table 3. Fitting degree of planning railway lines

Planning railroad	Design speed (Km/h)	Railroad length (Km)	Fitting degree
Hu-Kun	350	482.82	0.38
Cheng-Gui	250	174.44	0.29
Bijie-Xingyi	200	320.34	0.21
Liupanshui-Guiyang-Zunyi	250	409.27	4.24
Guiyang-Duyun	200	395.04	0.14
Guiyang-Kaili	200	120.96	0.17
Tongren-Yuping	200	65.94	0.01
Zhaotong-Jishou	160	510.94	0.01
Liupanshui-Jinzhi	120	89.06	0.41
Guiyang-Duyun-Enshi	250	90.60	3.20
Duyun-Hechi	160	88.49	0.06
Wengan-Fuquan	120	49.66	0.02
Xuyong-Zhijin-Tianlin	120	289.98	0.72
Jinsha-Xindian	120	58.16	0.10
Xingyi-Dushan	160	279.08	0.14
Chongqing-Guiyang	200	345.94	1.54
Qianjing-Huaihua	200	111.95	0.38
Dushan-Yongzhou	160	215.45	0.08
Zunyi-Wengan-Kaiyang	160	147.09	0.33
Huangtong-Zhijin	120	60.53	0.03
Gui-Zhu	300	261.93	0.15

3.2 Railway Protection Grade Assessment

Based on extension method, this paper selected three elements as the matter-element model: train speed, hazard weight and fitting degree. The specific steps were as follows [10–13]:

(1) Determining the classical domain

Table 4 lists the five protection grades according to the numerical characteristics of train speed, hazard weight and fitting degree, respectively.

(2) Determining the segment domain

The segment domain of the protection grade, namely the upper-lower limit of classical domain is expressed as:

$$\mathbf{R}_p = (\mathbf{N}, \mathbf{c}, \mathbf{v}) = \begin{bmatrix} \mathbf{N} & c_1 & V_{p1} \\ & c_2 & V_{p2} \\ & c_3 & V_{p3} \end{bmatrix} = \begin{bmatrix} \mathbf{N} & c_1 & \langle 120, 360 \rangle \\ & c_2 & \langle 0, 4.5 \rangle \\ & c_3 & \langle 1.0, 6.3 \rangle \end{bmatrix} \quad (1)$$

Table 4. The factors affecting the railway protection grades

Element index	Protection grade				
	I	II	III	IV	V
Train speed C_1	120–160	160–220	220–260	260–320	320–360
Fitting degree C_2	0–0.05	0.05–0.2	0.2–0.5	0.5–1.0	1.0–4.5
Hazard weight C_3	1.0–2.0	2.0–3.0	3.0–4.0	4.0–5.0	5.0–6.3

Where R_p means matter-element model of the railway protection grade; N refers to the protection grade of railway to be evaluated; c_i indicates the evaluation indices; V_{pi} is the value range of the evaluation indices.

(3) Determining the matter-element

Planning railway is expressed in the form of matter-element as:
Hu-Kun passenger special line:

$$\mathbf{R}_1 = \begin{bmatrix} \mathbf{P}_1 & c_1 & 350 \\ & c_2 & 0.38 \\ & c_3 & 4.69 \end{bmatrix} \tag{2}$$

Gui-Zhu fast railway:

$$\mathbf{R}_{21} = \begin{bmatrix} \mathbf{P}_{21} & c_1 & 350 \\ & c_2 & 0.15 \\ & c_3 & 2.96 \end{bmatrix} \tag{3}$$

(4) Determining the connection degree between matter-elements and protection grade

$$\rho(v_i, V_{ti}) = \left| v_i - \frac{a_i + b_i}{2} \right| - \frac{b_i - a_i}{2} \quad \text{(i = 1,2...n)} \tag{4}$$

$$\rho(v_i, V_{pi}) = \left| v_i - \frac{a_{pi} + b_{pi}}{2} \right| - v_i - \frac{b_{pi} - a_{pi}}{2} \quad \text{(i = 1,2...n)} \tag{5}$$

a_i and b_i are the upper or lower limit of V_{ti}; v_i is the value of the evaluation indices.
The connection function between i-th element and j-th protection grade is:

$$K_j(v_i) = \begin{cases} \frac{-\rho(v_i, V_{ti})}{|V_{ti}|}, & v_i \in V_{ti} \\ \frac{\rho(v_i, V_{ti})}{\rho(v_i, V_{pi}) - \rho(v_i, V_{ti})}, & v_i \notin V_{ti} \end{cases} \tag{6}$$

(5) Determining the weight coefficient

In this paper, the weight coefficient $\lambda_1 = \lambda_2 = \lambda_3 = 1/3$

(6) Calculating the connection degree based on connection function and weight coefficient

$$K_1(p) = \sum_{i=1}^{3} \lambda_i K_1(v_i) \tag{7}$$

Repeating steps (4), (5), (6), the connection degree under other protection grade can be obtained. $K_2(p)$, $K_3(p)$, $K_4(p)$, $K_5(p)$

$K_{j0} = \max_{j\in(1,2,\ldots,m)} K_j(v_i)$, the matter-element belongs to K_{j0}. The protection grades of the 21 planning railways are shown in Table 5.

Table 5. The protection grades of each railway

Railway lines	Protection grade					
	K1	K2	K3	K4	K5	Grade
Hu-Kun	−0.77	−0.70	−0.53	−0.25	−0.02	V
Cheng-Gui	−0.76	−0.68	−0.53	−0.20	−0.01	V
Bijie-Xingyi	−0.67	−0.55	−0.37	0.17	−0.19	IV
Liupanshui-Guiyang-Zunyi	−0.86	−0.82	−0.75	−0.53	0.11	V
Guiyang-Duyun	−0.55	−0.35	−0.12	0.06	−0.35	IV
Guiyang-Kaili	−0.61	−0.46	−0.27	0.28	−0.28	IV
Tongren-Yuping	0.5	−0.61	−0.50	0.06	−0.40	I
Zhaotong-Jishou	−0.41	0.26	−0.17	−0.16	−0.41	II
Liupanshui-Jinzhi	−0.43	−0.23	0.27	−0.10	−0.39	III
Guiyang-Duyun-Enshi	−0.57	−0.42	−0.31	−0.44	−0.40	III
Duyun-Hechi	−0.46	−0.26	0.10	−0.04	−0.35	III
Wengan-Fuquan	−0.20	−0.27	−0.04	−0.30	−0.53	III
Xuyong-Zhijin-Tianlin	−0.43	0.24	0.23	0.00	−0.36	II
Jinsha-Xindian	−0.65	−0.50	−0.27	−0.24	0.01	V
Xingyi-Dushan	−0.41	−0.16	0.30	−0.16	−0.41	III
Chongqing-Guiyang	−0.66	−0.49	−0.58	−0.55	−0.01	V
Qianjing-Huaihua	−0.34	0.17	−0.07	−0.38	−0.58	II
Dushan-Yongzhou	0.09	0.00	−0.50	−0.66	−0.75	I
Zunyi-Wengan-Kaiyang	0.10	−0.22	−0.46	−0.69	−0.80	I
Huangtong-Zhijin	−0.86	−0.81	−0.74	−0.44	−0.11	V
Gui-Zhu	−0.55	−0.30	−0.25	0.03	−0.38	IV

It can be seen from Table 5 that the overall 21 planning railways need moderate or high grade protection, and there are even many railroads need key protection such as the six railroads in V protection grade and the four railroads in IV protection grade. The protection grades of the 21 railways are shown in Fig. 6.

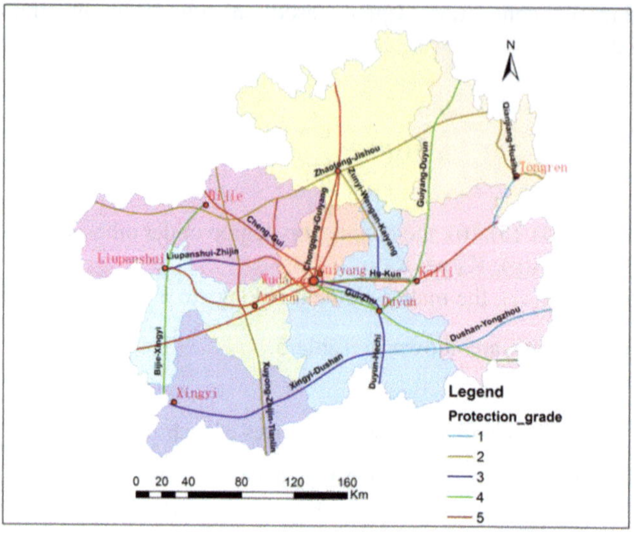

Fig. 6. The protection grade of the 21 railways in Guizhou Province

3.3 Verification of the New Evaluation Method

Among the twenty-one railways, the Hu-Kun railway was in V protection grade, which was the key section to protect. In recent years, Hu-Kun railway area has suffered large amount of geological disasters, and several railway interruption accidents have occurred in Guizhou Province and its surrounding areas. On June 18, 2015, Hu-Kun high-speed railway in Guizhou Province was affected by geological disasters, resulting in the suspension of six multiple unit trains. In July 2014, landslide occurred in the eastern section of Hu-Kun railway, causing 12 h delay of the train. On June 5, 2015, since the heavy rainfall induced landslides, during the railroad broken time, the railway station applied passengers for more than 120 thousand refund or changed tickets. Hu-Kun railway is one of the most serious railroads affected by geological disasters gravely in Guizhou Province. It poses a severe threat to the safety of people's lives and properties and it has an important impact on the traffic safety of Guizhou Province, which are consistent with the evaluation results.

4 Conclusions

1. Competition network model is used to divide the landslide risk mapping into 5 grades. With the increase of risk degree, the distribution of the historical landslide disaster is denser, which indicates that the competition network model is effective and sensitive for the landslide susceptibility mapping.
2. According to the distribution of landslide disaster points on both sides of railway, this paper puts forward the concept of "fitting degree". The greater the fitting degree

is, the closer the landslide disaster points are to the railway line, indicating that the risk of the railway is high, and the higher protection grade will be needed.

3. Extension method is used to calculate the protection grade for planning railway on geological disasters. As establishing the matter-element model, three elements are used: the train speed, fitting degree and the grade of geological disaster susceptibility.

Acknowledgement. The authors wish to acknowledge the support and motivation provided by Geological Hazard Susceptibility Mapping and Assessment in Basaltic Area (No: 2009318 802074), and West Project of Ministry Communication, China (N0. 2009318000074).

References

1. Varnes, D.J.: Slope Movement Types and Processes. Landslides: Analysis and Control. Transportation Research Board Special Report. National Academy of Sciences (1978)
2. Pistocchi, A., et.al.: The use of predictive modeling techniques for optimal exploitation of spatial databases: a case study in landslide hazard mapping with expert system-like methods. Environ. Geol. (2002). doi:10.1007/s002540100440, Springer
3. Yesilnacar, E., Hunter, G.J.: Application of Neural Networks for Landslide Susceptibility Mapping in Turkey. Kluwer Academic Publishers, Springer, The Netherlands, Dordrecht (2004). doi:10.1007/1-4020-2409-6_1
4. Melchiorre, C., et.al.: Artificial neural networks and cluster analysis in landslide susceptibility zonation. Geomorphology, ScienceDirect (2008). doi:10.1016/j.geomorph. 2006.10.035
5. Yin, K.L., et.al.: Early-warning and prediction of abrupt geological hazards in Zhejiang Province. China University of Geosciences Press (2005)
6. Dai, F.C., et.al.: Landslide risk assessment and management: an overview. Eng. Geol. ScienceDirect (2002). doi:10.1016/s0013-7952(01)00093-x
7. Zhou, M., Ren, P.A.: Fitting degree of regression straight line. J. Xi'an Eng. Univ. (1999)
8. Li, W.L., Cui, J.K.: On parameter estimation of the largest fittings. J. Shaanxi Inst. Technol. (2002)
9. Fang, H.Y., Chen, J.J.: BP model for hydrologic series prediction and goodness of fit analysis. J. Yangzhou Univ. (2001)
10. Li, Y.F., Liu, S.J.: The matter element analytical method of the assement of rock mass stability. Gold (1998)
11. Zuo, C.Q., Chen, J.P.: Rock mass classification based on extenics theory applied in metamorphic soft rock tunnel. Geol. Sci. Technol. Inf. (2007)
12. Hao, H.C., Zhu, F.H.: Recognition method of potential failure mode of slope based on extenics theory. Chin. J. Undergr. Space Eng. **4**, 024 (2007)
13. Wang, L.: Topology analysis for steadiness of side slope of rock body. Hebei Metall. **1**, 21–23 (1999)

Prediction of Travel Time Estimation Accuracy in Connected Vehicle Environments

Osama A. Osman(✉) and Sherif Ishak

Louisiana State University, Baton Rouge, LA, USA
othabel@lsu.edu

Abstract. Effective management of transportation networks requires accurate travel time information that is largely determined by the quality of collected real-time traffic data. In connected vehicle (CV) environments, wherein equipped vehicles may be the primary source of reliable travel time data, accuracy and reliability of travel time estimates present a challenge due to the low market penetration at the early deployment stages of CV technology. The absence of ground truth data presents another challenge for quantifying the accuracy and reliability of travel time estimates. Therefore, CV infrastructure should be well planned in transportation networks to achieve acceptable and reliable estimates of travel times. Recent research shows that the accuracy of travel time estimates is influenced by traffic density, CV market penetration, and transmission range. These factors also impact the vehicle-to-vehicle and vehicle-to-infrastructure communication stability in a transportation network. This suggests correlation between the accuracy and reliability of travel time estimates and the communication stability. This study develops regression models to measure the accuracy and reliability of travel time estimates as a function of communication stability. Such models can help transportation planners assess the anticipated accuracy and reliability of travel time estimates in CV environments, as well as make better infrastructure investment decisions to ensure an acceptable level of accuracy and reliability of travel time estimates.

1 Introduction

Connected vehicle is a rapidly emerging, promising, challenging, and self-organized communication technology that is taking place in the near future. It is aiming at using moving vehicles as basis for safety and comfort information transmission in order to have a highly operable and safe transportation environment. In connected vehicle environments, information is transmitted between nodes (vehicles) with high mobility and also limited degrees of freedom that is determined by the transportation network and the surrounding vehicles. Therefore, a highly efficient and dynamic broadcasting protocol that guarantees a high connectivity is required. As such, several studies have been conducted to develop an efficient protocol that can cope with the highly dynamic vehicular environment.

Measuring the performance of connected vehicle environments is a critical factor in its planning. Performance measures give an indication of the way connected vehicle environments will perform when specific market penetration, traffic density, and

J. Pombo and G. Jing (eds.), *Recent Developments in Railway Track and Transportation Engineering*, Sustainable Civil Infrastructures, DOI 10.1007/978-3-319-61627-8_6

transmission range values are available. One of the recently developed performance metrics is connectivity robustness (CONROB) [1]. CONROB is a network level metric that measures stability of existing communications between connected vehicles, as well as the potential of disconnected vehicles to connect and gain significant area coverage over the road network [1]. Connectivity robustness metric accounts for the geometric characteristics of the road network in addition to vehicular traffic related factors (traffic density and market penetration) as well as wireless communication related factors (transmission range).

Connected vehicle technology provides a probe vehicle feature that enables data collection about vehicle state. The collected data can be used to estimate traffic performance measures such as travel time which is considered an important piece of information for traffic management. Travel time estimates are critical for traffic management centers (TMC) to monitor traffic conditions in road networks and, consequently, make decisions aiming to improve the network operability. For this traffic management process to be efficient, accurate and reliable travel time estimates are required. In connected vehicle environments, accuracy and reliability of travel time estimates may not be accessible pieces of information, especially that connected vehicles may be the primary source of data. This in turn will lead to inability to judge the validity of travel time estimates and hence may convert the transportation network to a black box for officials at TMCs. In fact, travel time estimates accuracy and reliability are determined by the quality of collected data which is impacted by CV market penetration, traffic density, and CV transmission range. These factors also affect communication stability in connected vehicle environments, which offers correlation between communication stability and accuracy and reliability of travel time estimates. This correlation suggests using CONROB metric, as a measure for communication stability, in assessing the accuracy and reliability of travel time estimates in transportation networks. More specifically, CONROB metric, as a planning tool, can be used, during the planning stage of connected vehicle infrastructure, to assess the anticipated level of estimates accuracy and reliability that can be obtained from a transportation network. Consequently, if poor and/or unreliable estimates are anticipated, transportation planners can make decisions to provide and better allocate the required connected vehicle infrastructure to maintain a minimum level of data quality, which will help to achieve accurate and reliable travel time estimates. As such, this paper develops regression models to evaluate the accuracy and reliability of travel time estimates obtained from a connected vehicle environment with specific system characteristics (market penetration, transmission range, and traffic density) as a function of CONROB metric.

2 Background

2.1 Performance Measures for Connected Vehicle Environments

There has been a significant effort to come up with some performance metrics for connected vehicle environments [2–11]. These studies are concerned about the connectivity probability, the healing time after a connectivity failure, the delay in message delivery, the vehicular traffic delay time, or other vehicular traffic metrics. Spanos and Murray [12] developed the geometric connectivity robustness model to measure the connectivity robustness between connected vehicles in terms of their freedom of movement relative to each other. This model accounts for the transmission range and the spatial distribution of vehicles relative to each other in the network. However, it does not account for critical factors such as market penetration, traffic density, and the spatial propagation of the wireless signal over the road network. This was overcome in the CONROB model that was developed by Osman and Ishak [1]. CONROB model is a network level metric that measures the stability of existing communications between connected vehicles as well as the potential of the disconnected vehicles to reconnect again with better ability for information to propagate over significant distances in the road network. CONROB is based on Newton's universal law of gravitation and accounts for multiple factors describing the system characteristics of connected vehicle environments such as market penetration, wireless transmission range, spatial distribution of vehicles relative to each other, the spatial propagation of the wireless signal, and traffic density. Simulation runs were conducted to verify the developed model. The results provide an evidence of the ability of the model to capture the effect of the different factors on the connectivity between vehicles, which provides a viable tool for assessing connected vehicle environments.

2.2 SAE-J2735 Protocol

The equipped vehicles in connected vehicle environments collect data in the form of data snapshots that provide information about the vehicle state such as the vehicle coordinates and speed. The data snapshots are collected and stored on the vehicle's On-Board Units (OBUs). These snapshots are then transmitted to the first RSU the vehicle passes by through the V2I application of the connected vehicle technology. The data snapshots are collected according to the Society of Automotive Engineers (SAE) J2735 standard. The SAE-J2735 standard allows snapshots collection on a vehicle's OBU at predefined time intervals according to the vehicle speed. In order to have a reasonable representation to the vehicle movement, the snapshots collection time interval decreases as the vehicle speed decreases. In other words, the SAE-J2735 standard allows periodic snapshots to be collected every 20 s as the vehicle speed is higher than 60 mph, every 4 s as the vehicle speed is below 20 mph, and at a linearly interpolated time intervals when the vehicle speed is between 20 and 60 mph.

The collected snapshots in the SAE-J2735 standard represent two main data types: Probe Data Messages (PDMs) and Basic Safety Messages (BSMs). The information provided in the PDMs only include vehicle locations and speeds. Whereas, the BSMs include, in addition to the vehicle speeds and locations, braking state, emergency braking instances, and acceleration and deceleration values.

While the SAE-J2735 standard is the basic data snapshots collection protocol proposed for the connected vehicle technology, there were some attempts to come up with a more effective protocol. For instance, Chen et al. [13] presented a new protocol called R2 protocol for collecting probe data in connected vehicle environments. The protocol is based upon collecting a vehicle snapshot when a significant change in the vehicle speed happens. Two data sets obtained from a connected vehicle simulation test bed and a real world test bed in Michigan were used to evaluate the R2 protocol. The proposed protocol was compared to three existing protocols (Fixed 2 s, Fixed 4 s, and SAE J2735 protocols). The results show that the R2 protocol outperforms the other three protocols in terms of the error measure in the collected data with fewer snapshots than those required for the other three protocols.

2.3 Traffic Conditions Estimation

The data collection capability of connected vehicle technology provides (the probe vehicle feature) can be used to estimate vehicular traffic measures of effectiveness (MOEs) such as queue length, speed, and travel time. Several studies investigated the probe vehicle feature of the connected vehicle technology. For instance, Comert and Cetin [14] studied the effect of market penetration on the estimation accuracy of queue lengths. In their study, the authors were able to identify the queue length using only the location information of the last equipped vehicle in the queue. Osman et al. [15] studied the effect of traffic stochastic nature on queue length estimation accuracy. In their study, the authors tested the effect of several factors on queue length estimation accuracy such as, traffic density, vehicle arrival, and market penetration. The authors concluded that queue length estimation in connected vehicle environments in controlled by an index (queue formation coverage index) that defines the stochastic nature of traffic at intersections. Arogate et al. [16] presented estimation methods for different traffic MOEs at arterial as well as intersection level such as average speed, acceleration noise, and queue length. The authors tested their estimation methods at undersaturated and saturated traffic conditions and determined the required market penetration for accurate estimation of the MOE's. They concluded that for saturated traffic conditions, the required market penetration for accurate estimate of the different MOE's is less than that is required for the undersaturated conditions. More specifically, they concluded that higher accuracy can be obtained in oversaturated traffic conditions even at the early stages of connected vehicle deployment when the market penetration rates are low. The common factor in these studies was the accuracy of the obtained estimates, which is determined by the amount of data available and how much it covers of the traffic situations in the road network. The main factors affecting the accuracy are the effective communication range, penetration rate, snapshots resolution, buffer size of the OBUs, and number of active RSUs.

The probe vehicle feature of connected vehicle technology enables real-time data collection which in turn enables real-time estimation of MOEs. Past research showed that the accuracy of these estimates is affected by many factors such as the transmission range, market penetration, and the traffic density. Since these factors also affect communication stability, relationships between communication stability and MOEs estimation accuracy and reliability could be developed. This paper investigates the correlation between the connectivity robustness metric, as a measure of communication stability, and the accuracy and reliability of traffic conditions estimates. Then, relationships are established to measure the accuracy and reliability of MOEs estimates, obtained from a connected vehicle environment with specific system characteristics, as a function of communication stability. More specifically, this paper focuses on measuring travel time estimation accuracy as well as estimation reliability as functions of the connectivity robustness metric.

3 Connectivity Robustness Model (CONROB)

Connectivity robustness is a measure for the stability of existing communications between connected vehicles. In addition, it gives indication for the potential of the vehicles that are not able to communicate with each other to connect to their neighboring vehicles and merge into a cluster covering significant area of the road network [1]. Basically, because of the dynamic nature of traffic and the effect of the penetration rates, not all equipped vehicles will be able to communicate with each other and hence vehicles falling in the transmission range of each other form clusters that have different sizes measured by the number of vehicles contained and the area covered by that cluster. CONROB model treats clusters of connected vehicles as masses and measures the attraction force between them in an analogy to the Newton's law of universal gravitation. Cluster mass is measured as a function of the cluster size and the attraction force is the connectivity robustness. Connectivity robustness is also measured as a function of the distance between clusters relative to the transmission range of the vehicles. More specifically, the model accounts for connected vehicle penetration rate, transmission range, communication area coverage, which is measured by the area of clusters in which a piece of information could be exchanged between all vehicles in the cluster, spatial distribution of connected vehicles relative to one another, and traffic density. As shown in Eq. (1), the model deals with two clusters of vehicles at a time, where the cluster is a group of connected vehicles (traveling in both directions) that are within the transmission range of one another.

$$R_{n,m}(t) = \frac{\frac{4}{N_C(t)} \cdot \left(\frac{R_{l,n}(t)}{\min\limits_{i \in N_n}\{r_i(t)\}} \cdot \left(\frac{N_n(t)}{N_T(t)}\right)^{0.5} \cdot \left(\frac{a_n(t)}{A}\right)^{0.5} \right) \cdot \left(\frac{R_{l,m}(t)}{\min\limits_{i \in N_m}\{r_i(t)\}} \cdot \left(\frac{N_m(t)}{N_T(t)}\right)^{0.5} \cdot \left(\frac{a_{m(t)}}{A}\right)^{0.5} \right)}{\left(\frac{\min\limits_{i \in N_n \& j \in N_m} d_{i,j}(t)}{\min\limits_{i \in N_n \& j \in N_m} \{r_i(t), r_j(t)\}} + 1 \right)^2} \quad (1)$$

where, $R_{n,m}(t)$ is the connectivity robustness between clusters n and m at time t, $N_C(t)$ is the number of probe vehicle clusters in the road network, $R_{l,n}(t)$ is the geometric robustness of cluster n, discussed in details in [1], $N_n(t)$ is the number of vehicles in cluster n, N_T is the total number of vehicles in the road network, $a_n(t)$ is the area covered by the wireless signal of vehicles in cluster n, A is the total area of the road network, $d_{(i,n),(j,m)}(t)$ is the distance between vehicle i in cluster n and vehicle j in cluster m, and $r_{i,n}(t)$ is the transmission range of vehicle i in cluster n.

Equation 1 is used for every two clusters in the road network, and then the network robustness, which is the average robustness over the network at time t, is calculated as in (2).

$$R_{\text{network}}(t) = \frac{1}{N_C(t)} \cdot \sum\nolimits_{n=1}^{N_C(t)} R_{n,m}(t) \tag{2}$$

Because the network robustness is time dependent, it is defined as a distribution that is represented by the average robustness value over the total time (*T*) during which connectivity robustness is calculated. This is referred to as the overall robustness and is defined by (3).

$$R_{overall} = \frac{1}{T}\sum\nolimits_{t=0}^{T} R_{network}(t) \tag{3}$$

Further discussion for the mathematical derivation of the model was presented in the authors' previous study [1]. Connectivity robustness, as a metric for communication stability in connected/automated vehicle environments, accounts for multiple factors that make it able to capture the geometric, communication, and traffic characteristics in a large scale road network.

4 Methodology

The accuracy of estimated traffic information by connected-vehicle collected data is totally dependent on the factors affecting the connectivity robustness such as transmission range, market penetration, traffic density, spatial distribution of vehicles and spatial propagation of vehicles. This suggests that a relationship between connectivity robustness and the accuracy of transmitted traffic information exists. As such, this relationship between connectivity robustness and the accuracy of estimated travel time information in connected vehicle environments is investigated. More specifically, two relationships are investigated: the connectivity robustness with the travel time estimation error and the connectivity robustness with the standard deviation of the travel time estimation error. The standard deviation is used as a measure for the reliability of the travel time estimation.

4.1 Simulation Runs

The aforementioned relationships are investigated using simulation results of, a total of 10 simulation runs for each of 12 different scenarios (4 traffic density values X 4 market penetration values) obtained from a 500 sq-km network shown in Fig. 1. The traffic volumes in the network links, provided and calibrated by PTV, were multiplied by a factor of 0.5, 0.75, 1, and 1.25, in order to generate four different flow scenarios. Each flow value was translated into a traffic density normalized by the network area (a number of vehicles per sq-km) based on the results of the simulation runs (60, 90, 130, and 190 vehicles/sq-km). These values were selected in order to have reasonably free flowing traffic. For the market penetration, three values of 5%, 15%, 25%, and 50% were considered and assumed to be consistent on all the network links in the beginning of each simulation run.

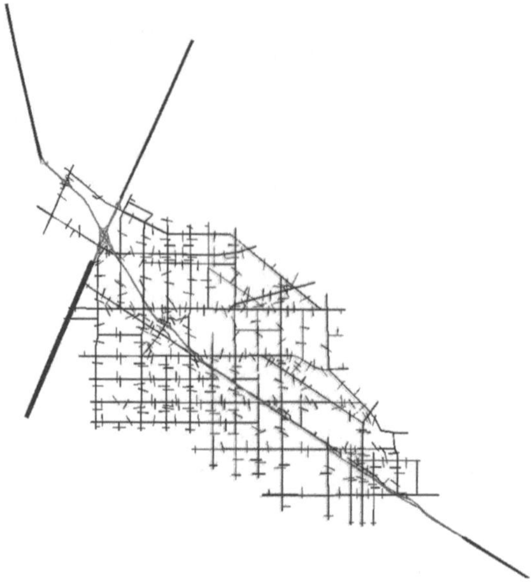

Fig. 1. Study road network

The simulation runs were performed considering dynamic transmission range to account for the randomness effect resulting from the surrounding obstacles. The actual transmission range is calculated, according to Arogate et al. [16], as a ratio of the maximum transmission range of the wireless channel as in (4).

$$R = T_r \cdot (1 - \epsilon); 0 < \epsilon \leq 1 \tag{4}$$

where T_r is the maximum transmission range of the wireless channel, and the parameter ϵ refers to the wireless channel fading conditions at the vehicle position. This simplified equation reflects the practical transmission range (R) that is determined based upon the surrounding conditions to the vehicle. This is determined using the parameter ϵ which changes according to the area type (downtown or suburbs). In downtown areas, for instance, the high rise buildings, industries, and other installations, will interfere with the wireless signals of the vehicles. In such case, the value of ϵ will be high so that the practical transmission range might be less than the maximum value. On the other hand, in the suburbs, the obstacles are fewer and hence the interference with the wireless signals is much less, resulting in a small value of ϵ. In such case the value of the practical transmission range is very close to the maximum value [17].

In order to apply this dynamic transmission range equation in the simulation, the selected network is divided into four zones with almost similar characteristics in each zone. For each zone, an arbitrary range is selected for the fading error according to the prevailing conditions. Then, based on these arbitrary ranges and a maximum transmission range value of 1000 m, the actual transmission range for each 3 m × 3 m area block is calculated. This is performed using a MATLAB code executed for each simulation time step.

For the travel time estimations, a number of RSUs, each with a 150 m transmission range, are placed across the entire road networks. The number and locations of the RSUs are determined using MATLAB based on a proposed deployment plan that suggests every RSU to cover an area of the network determined by a radius of 365 m. Based on this deployment plan, 46 RSU are placed at 46 intersections distributed uniformly (with uniform separating distances between one another) over the road network. PDMs are then collected by connected vehicles for each simulation time step according to the SAE-J2735 protocol. This was performed using the open access Trajectory Conversion Algorithm (TCA) software that integrates VISSIM-C2X feature with Python programming interface.

4.2 Travel Time Estimation and Estimation Error

Using the speed information provided by the collected data snapshots, the travel time information is estimated for each of 193 links with different lengths using the approach implemented by Kianfar ane Edara [18], as in Eq. (5). The selected links are distributed over the network area to give a reasonable representation about the traffic conditions in the entire network. On the other hand, the speed data provided by all vehicles in VISSIM output on each of the selected links are used to calculate the ground truth travel-time information. The travel time as well as the connectivity robustness are calculated over 12 time steps for each simulation run.

$$\mathrm{TTE}_{l,t} = \frac{\mathrm{L}_l}{\sum_{s=1}^{S(t)} \sum_{n=1}^{N(t)} \alpha_{s,t} \cdot \beta_s \cdot (v_{s,n,t} \cdot p_{n,l,s,t}) / \sum_{s=1}^{S(t)} \sum_{n=1}^{N(t)} \alpha_{s,t} \cdot \beta_s \cdot p_{n,l,s,t}} \tag{5}$$

Where $TTE_{l,t}$ is the estimated travel time for link l at time step t, L_l is the length of link l, $v_{s,n,t}$ is the speed information of connected vehicle n obtained from snapshot s at time step t, $\alpha_{s,t}$ is a binary factor (0 or 1) that determines whether a snapshot is transmitted to an RSU at time step t, β_s is a binary factor that determines whether a snapshot is collected on the OBU at time step t, $p_{n,l,s,t}$ is a binary factor that determines if a snapshot represents the vehicle state while it is on link l, $N(t)$ is the total number of connected vehicles in the network at time step t, and $S(t)$ is the total number of snapshots transmitted to the RSUs at time step t. For a snapshot to be considered in the travel time estimation equation, it needs to be collected on the link of interest and transmitted to a RSU within the time step of interest.

The absolute percentage travel-time-estimation-error (*TTEE*) is then calculated as an average value for the 193 links over the simulation time period *T*, as in (6).

$$\mathrm{TTEE} = \frac{1}{\mathrm{T}} \cdot \sum_{t=1}^{\mathrm{T}} \frac{1}{\mathrm{L}} \cdot \sum_{l=1}^{\mathrm{L}} \left| \frac{\mathrm{TTE}_{l,t} - \mathrm{TTG}_{l,t}}{\mathrm{TTG}_{l,t}} \right| \times 100 \tag{6}$$

5 Analysis of Results

5.1 Travel Time Estimation Error

The accuracy of travel time estimates is totally dependent on the available PDMs from the network which is controlled by the available number of connected vehicles. As the market penetration increases, more connected become available in the network which can help improve the accuracy of travel time estimates. For a pre-specified market penetration, due to the vehicles' distribution over the network, route choice decisions, and distribution of vehicle arrivals, every link in the network may have a number of connected vehicles that is either lower or higher than the design market penetration. As such, the average market penetration in the network turns out to be less than the design value, as shown in Fig. 2. The figure shows that the actual market penetration in the network changes between 71% and 77% of the design market penetration. The figure also shows that the actual market penetration becomes closer to the design value at higher traffic density values. Higher traffic densities give the opportunity for more uniform distribution of connected vehicles over the network, which allows the average market penetration in the network to be closer to the design value.

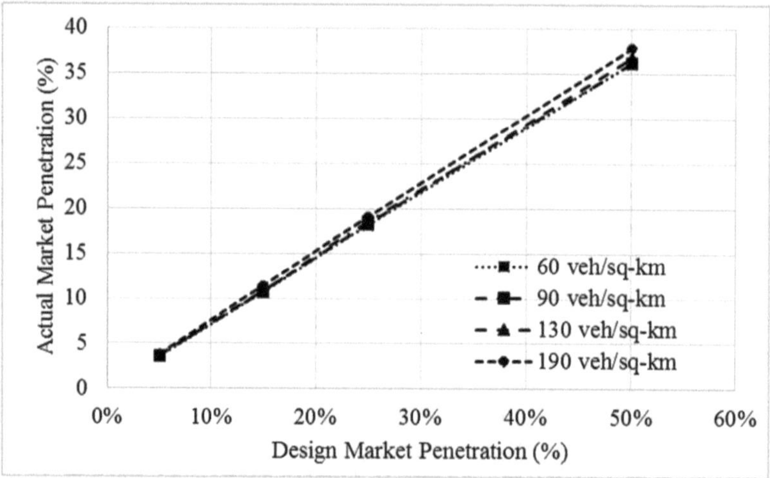

Fig. 2. Actual market penetration in the road network

The aforementioned variation in market penetration is determined by vehicles' distribution over the road links in the network. Some links may have very low number of connected vehicles or even no connected vehicles, as illustrated in Fig. 3a for a traffic density of 60 veh/sq-km. The figure shows that the percentage of links in the network travelled-over by mixed traffic with connected vehicles increases at higher market penetration values. As the market penetration increases, more connected vehicles are distributed over the network which reduces the percentage of links with no connected vehicles. More so, at higher market penetrations, vehicles' distribution over the network becomes more stable which results in reducing the variance of the percentage of links travelled-over by connected vehicles, as shown in the figure. This discussion applies to the different traffic density scenarios as shown in Fig. 3b, c and d.

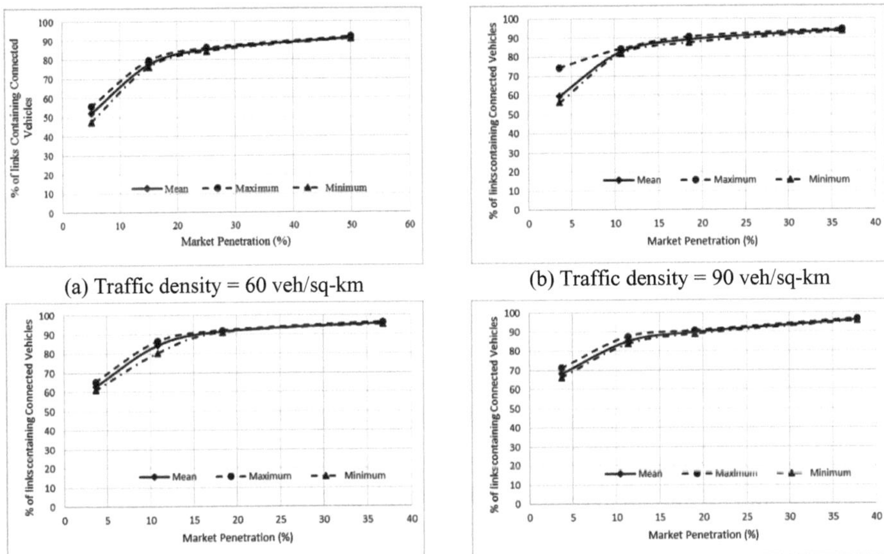

(a) Traffic density = 60 veh/sq-km

(b) Traffic density = 90 veh/sq-km

(c) Traffic density = 130 veh/sq-km

(d) Traffic density = 190 veh/sq-km

Fig. 3. Percentage links covered by connected vehicles

The previous discussion shows that higher market penetrations can provide better network coverage with connected vehicles. As the number of connected vehicles in the network increases, the number of road links travelled-over by connected vehicles increases. Thereby, larger number of PDMs can be collected that can cover larger areas of the road network and provide a better representation to the traffic conditions in the network. This is illustrated by the TTEE values in Fig. 4. The figure shows that, at higher market penetrations, more accurate travel time estimates are obtained. The figure also shows that the accuracy of travel time estimates deteriorates with the higher traffic densities, especially for market penetration values up to 25%. Whereas, as the market penetration increases to 50%, an infliction point takes place on the TTEE curve. The accuracy of the travel time estimates improves as the traffic density increases up to that infliction point, then the accuracy deteriorates as the traffic density increases beyond the infliction point. These results suggest that although the number of connected vehicles increases for the higher traffic densities, higher market penetrations are yet required to obtain accurate travel time estimates. This can be explained in light of traffic variability and the randomness. As the traffic density in the network increases, vehicles' distribution in the network changes randomly (only restricted by network geometry and enforced traffic rules). This may exasperate the variability of traffic conditions over the network which could lead to unreliable travel time estimates. In order to overcome such

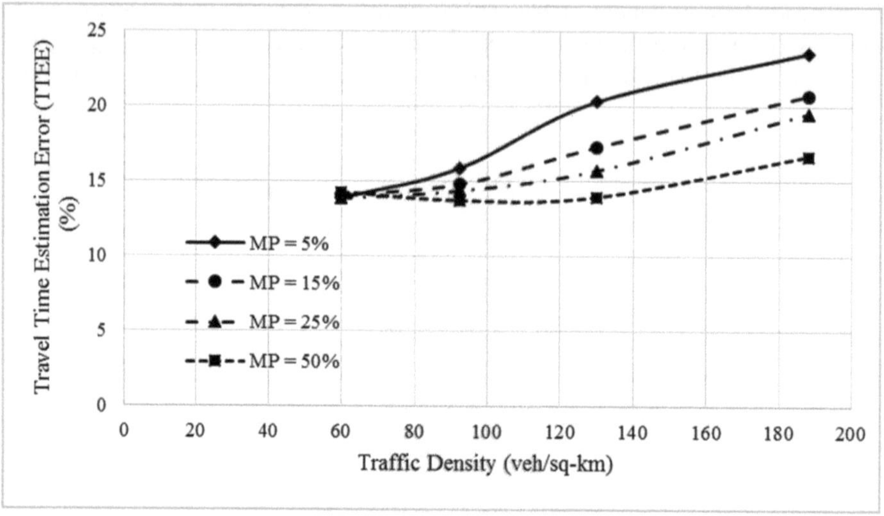

Fig. 4. Travel time estimation error (TTEE)

variability, higher market penetrations are required to allow more comprehensive data to be collected which can help better estimate traffic conditions in the network.

5.2 TTEE vs. Connectivity Robustness

The discussion in the previous section shows that TTEE depends on traffic density, market penetration, and distribution of connected vehicles in the network. These factors also affect the network connectivity robustness and stability, as discussed earlier. This indicates that the TTEE and the connectivity robustness could be correlated which is investigated in this section. Then, if such a correlation exists, relationships between TTEE and connectivity robustness is investigated.

Figure 5a illustrates the change in connectivity robustness as the market penetration and traffic density change. For market penetration values up to 25%, the network has low connectivity robustness values that do not change significantly as traffic density increases. This suggests that the available number of connected vehicles is not enough to reach a network coverage that can achieve higher connectivity robustness values. As the market penetration increases to 50%, the network coverage improves which results in a significant increase in the connectivity robustness. In such a case, an infliction point is clear in the trend of the connectivity robustness over the traffic density. The connectivity robustness changes significantly up to a 130 veh/sq-km density value. As the traffic density increases beyond that value, the change in the connectivity robustness is not as significant. These results indicate that, for the higher traffic density scenarios, higher market penetrations are required to achieve significant improvements in the connectivity robustness. This behaviour is consistent with the TTEE results in Fig. 5b which is discussed in the previous section.

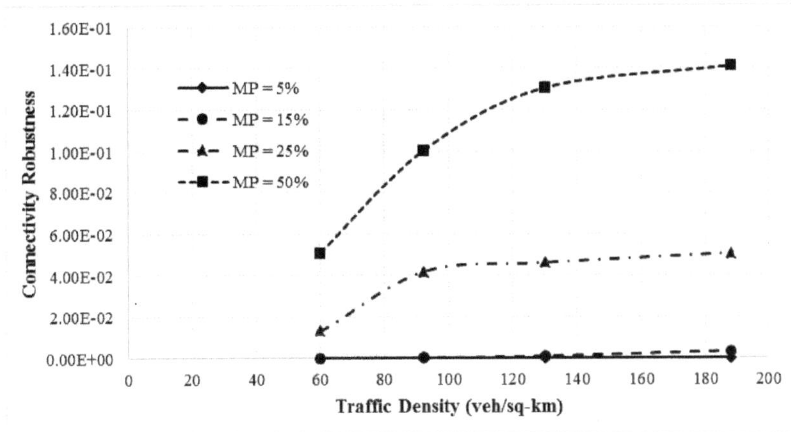

(a) Connectivity robustness

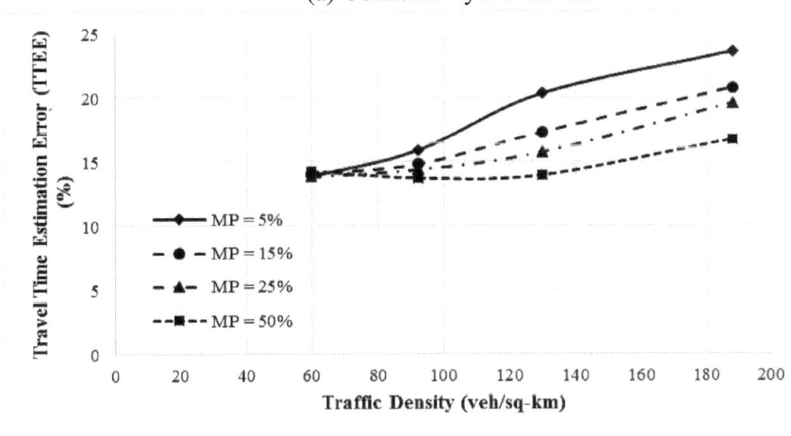

(b) Travel time estimation error

Fig. 5. Connectivity robustness vs. TTEE

Similar results are found for the standard deviation of the TTEE as a measure of the estimation reliability. Figure 6 shows that, for market penetration values up to 25%, no clear trend can be captured for the reliability of TTE. This may be related to the high variability in the traffic conditions. This variability, as discussed in the previous section, requires higher market penetration values to be overcome; this is the case for market penetration value of 50% as shown in the figure. In such a case, the standard deviation of TTEE decreases significantly as the traffic density increases up to 130 veh/sq-km (the inflection point in Fig. 5a). As the traffic density increases beyond that value, the decrease in standard deviation is not as significant. This behaviour is consistent with the connectivity robustness trend in Fig. 5a.

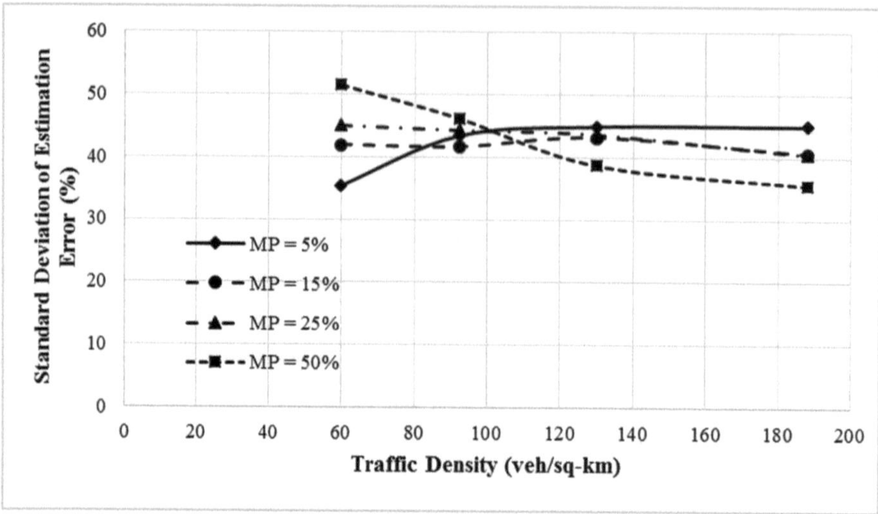

Fig. 6. Travel time estimation reliability

5.3 TTEE Accuracy and Reliability Regression Models

The previous discussion shows that the accuracy and reliability of travel time estimates are correlated with the connectivity robustness. In order to further investigate this, regression analysis is performed between the connectivity robustness and the TTEE as well as between the connectivity robustness and the standard deviation of the TTEE. The results of linear regression analysis are shown in Table 1. The table shows that the adjusted R^2 value for the regression model between the TTEE as the dependent variable and the logarithm of the connectivity robustness as the independent variable is 98% with a p-value of less than 0.0001. Similarly, the adjusted R^2 value for the model between the standard deviation of the TTEE and the logarithm of the connectivity robustness is 88% with a p-value of less than 0.0001. This implies that both the TTEE and the standard deviation of the TTEE are significantly sensitive to the connectivity robustness in the road network at 0.05 level of significance.

Table 1. Regression models of travel-time estimates accuracy and reliability

Measure	Model	Adj R-Sq	P-value
Travel time accuracy	TTEE = −4.003 + 0.229*log(CONROB)	98%	<0.0001
Travel time estimation-reliability	TTEE-St. Dev. = 9.458–3.134*log (CONROB)	88%	<0.0001

The developed regression models in Table 1 show that as the connectivity robustness increases, the TTEE and the standard deviation of TTEE decrease. This indicates that more stable and robust communication can help improve the quality of collected data from the transportation network, and hence help provide more accurate and reliable travel time estimates. In connected vehicle environments, where connected vehicles are expected to be the primary source of data, accuracy and reliability of travel time estimates may not be attainable pieces of information. Thus, the proposed models can, instead, be used during the planning stage of connected vehicle infrastructure to measure the accuracy and reliability of travel time estimates expected to be obtained from a transportation network with specific system characteristics. Consequently, transportation planners can determine and allocate the required connected vehicle infrastructure to provide a minimum level of connectivity and hence achieve an acceptable level of reliability and accuracy for travel time estimates.

6 Conclusions

Connected vehicle technology provides real time probe vehicle data that can be effectively used for estimation of traffic conditions in transportation networks. The accuracy of estimates is determined by the quality of collected data which is impacted by different factors such as the market penetration, traffic density, and transmission range. Such factors also influence the communication stability in connected vehicle environments. In this paper, the previously developed CONROB model was applied to assess the accuracy of travel time estimates in the network. This was investigated using results obtained from simulating a 500 sq-km road network in Washington County to the west of Portland, Oregon. Different MATLAB codes were developed to calculate the connectivity robustness and estimate travel times in the network over different time steps.

The simulation results showed that the accuracy of the travel time estimates deteriorated with higher traffic densities, especially for market penetration values up to 25%. Whereas the market penetration increased to 50%, the accuracy improved over the different traffic density scenarios. However, at very high density values, the results show slight deterioration in accuracy. Similarly, the reliability of travel time estimates did not show a clear trend for market penetration values up to 25%. As the market penetration increased to 50%, the travel time estimates became more reliable regardless of traffic density scenarios.

The connectivity robustness results showed similar performance to that of the travel time estimates. For market penetration values up to 25%, the network exhibited low connectivity robustness values that did not change significantly with the traffic density. As the market penetration increased to 50%, the network connectivity robustness improved significantly. These results indicated that the accuracy and reliability of travel time estimates are correlated with the network connectivity robustness. To confirm this observation, regression analysis was performed between the connectivity robustness and the accuracy of travel time estimates, as well as between the connectivity robustness and the reliability of travel time estimates. The results showed that the adjusted R^2 value for the regression model between the accuracy of travel time estimates and the logarithm of connectivity robustness was 98% with a p-value less than 0.0001. Similarly, the adjusted R^2 value of the relationship between the reliability of travel time estimates and the logarithm of connectivity robustness was 88% with a p-value less than 0.0001. This confirms that both regression models can be used as a planning tool to measure and help improve the level of accuracy and reliability of travel time estimates in transportation networks with connected vehicles infrastructure.

References

1. Osman, O., Ishak, S.: Accounting for traffic density and market penettration in a newly developed connectivity robustness model for connected vehicles environments. Presented at 93rd Annual Meeting of the Transportation Research Board, Washington, D.C. (2014)
2. Yousefi, S., Altman, E., El-Azouzi, R., Fahy, M.: Analytical model for connectivity in vehicular ad hoc networks. IEEE Trans. Veh. Technol. **57**(6), 3341–3356 (2008)
3. Panichpapiboon, S., Pattara-Atikom, W.: Connectivity requirements for self-organizing traffic information systems. IEEE Trans. Veh. Technol. **57**(6), 3333–3340 (2008)
4. Kafsi, M., Papadimitratos, P., Dousse, O., Alpcan, T., Hubaux, J.P.: VANET connectivity analysis. Presented at IEEE Workshop on Automotive Networking and Applications (2008)
5. Yousefi, S., Fathy, M.: Metrics for performance evaluation of safety applications in vehicular ad hoc networks. Transport **23**(4), 291–298 (2008)
6. Wisitpongphan, N., Bai, F., Mudalige, P., Tonguz, O.K.: On the routing problem in disconnected vehicular ad-hoc networks. Presented at 26th IEEE International Conference on Computer Communications, INFOCOM 2007 (2007)
7. Manvi, S.S., Kakkasageri, M.S., Mahapurush, C.V.: Performance analysis of AODV, DSR, and swarm intelligence routing protocols in vehicular ad hoc network environment. Presented at International Conference on Future Computer and Communication, ICFCC (2009)
8. El-atty, S.M.A., Stamatiou, G.K.: Performance analysis of Multihop connectivity in VANET. Presented at the 7th International Symposium on Wireless Communication Systems (ISWCS) (2010)
9. Reinders, R., Van Eenennaam, M., Karagiannis, G., Heijenk, G.: Contention window analysis for beaconing in VANETs. Presented at the 7th International Wireless Communications and Mobile Computing Conference (IWCMC) (2011)
10. Yang, Y., Bagrodia, R.: Evaluation of VANET-based advanced intelligent transportation systems. Proceedings of the Sixth ACM International Workshop on Vehicular Internetworking, pp. 3–12 (2009)
11. Roess, R.P., Prassas, E.S., McShane, W.R.: Traffic Engineering, 3rd edn. Prentice-Hall, Englewood Cliffs (2004)
12. Spanos, D.P., Murray, R.M.: Robust connectivity of networked vehicles. Presented at 43rd IEEE Conference on Decision and Control (2004)
13. Chen, C., Kinafar, J., Edara, P.: New snapshot generation protocol for travel time estimation in a connected vehicle environment. Presented at 93rd Annual Meeting of the Transportation Research Board, Washington, D.C (2014)
14. Comert, G., Cetin, M.: queue length estimation from probe vehicle location and the impact of sample size. Eur. J. Oper. Res. **197**(1), 196–202 (2009)
15. Osman, O.A., Bakhit, P.R., Ishak, S.: Queue estimation at signalized intersections using basic safety messages in connected vehicle environments. Paper presented at the Transportation Research Board 95th Annual Meeting (2016)

16. Arogate, J., Christofa, E., Xuan, Y., Skabardonis, A.: Estimation of arterial measures of effectiveness with connected vehicle data. Presented at 91st Annual Meeting of the Transportation Research Board, Washington, D.C (2012)
17. Taleb, T., Benslimane, A., Ben Letaief, K.: Toward an effective risk-conscious and collaborative vehicular collision avoidance system. IEEE Trans. Veh. Technol. **59**(3), 1474–1486 (2010)
18. Kinafar, J., Edara, P.: Placement of Roadside Equipment (RSE) in connected vehicle environment for travel time estimation. Presented at 92nd Annual Meeting of the Transportation Research Board, Washington, D.C (2013)

Application of Traffic Conflict Techniques as Surrogate Safety Measures: A Sustainable Solution for Developing Countries

S.M. Sohel Mahmud[1(✉)], Luis Ferreira[1], Shamsul Hoque[2], and Ahmad Tavassoli[1]

[1] School of Civil Engineering, The University of Queensland (UQ), Brisbane, QLD, Australia
smsohelmahmud@gmail.com
[2] Department Civil Engineering, Bangladesh University of Engineering and Technology (BUET), Dhaka 1000, Bangladesh

Abstract. Social, economic and infrastructure losses due to road traffic accidents and their consequences are very significant all over the world, particularly in developing countries. The evaluation of causative factors of accidents and the selection of remedial measures continues to be based mainly on traditional approaches. Whereas, accident statistics are frequently questioned due to large underreporting of accidents, injuries and property damages, coupled with incomplete and inconsistent recording of information on reported accidents. Poor timelines, ethical issues, biasness and human error are also critical issues. This paper present a comprehensive assessment of the data quality of reported accident databases, in terms of the degree and diversity of the reporting and recording inconsistency, using a case study from Bangladesh.

For a more rigorous and sustainable form of safety analysis there is a need for robust methods that may yield targeted safety measures without the need to use accident data. Application of traffic conflict techniques for the diagnosis of accidents has gained research interest as a proactive surrogate approach. However, this has been developed and tested primarily based on lane based homogeneous traffic conditions prevailing in developed countries. Development of advanced image processing systems, as well as video analysis techniques for automatic discrimination of conflicts, has open new prospects. Traffic safety micro-simulation modeling using surrogate indicators is also a promising advancement in this context. This paper provides a framework for safety evaluation beyond the traditional approaches with the integration of recent advancement in surrogate safety evaluation for non-lane based traffic environments. Finally, future research directions, designed to achieve sustainable road safety objectives in developing counties, are outlined.

J. Pombo and G. Jing (eds.), *Recent Developments in Railway Track and Transportation Engineering*, Sustainable Civil Infrastructures, DOI 10.1007/978-3-319-61627-8_7

1 Introduction

Road traffic accidents and their consequences are disproportionately higher in developing countries (WHO 2013, 2015). Safety assessment, selection of intervention and evaluation of treatments is mainly based on the application of traditional approaches, which rely mainly on the use of historical data. Such approach tend to be unreliable due to significant shortcomings associated with accident data reporting and recording (Archer 2005; OECD 1997; Yang 2012). In addition, this is a reactive approach which does not present the causative factors of accidents. This problem is even more severe in developing countries. Therefore; there is significant need for alternative safety evaluation approaches. One of the major contributions of this paper is a comprehensive assessment of the data quality of reported accident databases, highlighting the need for change in the methodological approach to ensure sustainability regarding traffic safety improving initiatives in developing counties. The paper presents a case study from Bangladesh regarding data quality issues. The level of underreporting has been evaluated using a comparative study between official reports and other sources of data collected from different sources, including studies conducted by different local and international agencies. Despite having a long history, the application of proximal surrogate measures, such as traffic conflict techniques for safety evaluation in non-lane based heterogeneous traffic environments, is almost absent in literature (Mahmud et al. 2016). The main contribution this paper is a framework for proactive surrogate safety evaluation and management, which goes beyond traditional approaches for developing counties where traffic is heterogeneous and lane-discipline is poor. Finally, directions for future research have been put forwarded to make the system more compatible and reliable for developing countries traffic environments, as well as to address some research gaps for achieving sustainable road safety objectives in developing counties.

In this study, relevant articles were identified from various online and offline databases including the Google Scholar, Research Gate, Science Direct and Transport Research International Documentation, Engineering Village and other. Principles, important features, advantages and disadvantages of each traditional approach were documented through a thorough literature review. Crash data has been collected from the Accident Research Institute (ARI), Bangladesh Police and Bangladesh Bridge Authority (BBA). World-wide reports on road traffic accident statistics published by WHO have been used.

The paper has been organized as follows. Firstly, a summary of the traditional approaches for traffic safety evaluation is provided. The need for alternative safety evaluation approaches is then assessed and a case study is put forward. The assessment mainly consists of two major sub-sections, namely underreporting of accidents and consistency and completeness of reporting. This is followed by the framework of alternative approaches for safety evaluation in developing countries. Finally, a brief summary is provided followed by avenues of further research.

2 Traditional Approaches for Traffic Safety Evaluation (TSE)

Most of the current literature covering highway safety analysis, or traffic safety risk evaluation, has been conducted using traditional methods, particularly in the case of developing countries (St-Aubin 2011). These traditional approaches are mainly based on observed historical accident data using various types of statistical methods and anticipatory estimation studies based on safety audits (Roper and Turner 2008; St-Aubin 2012; Yang 2012). A list of those widely used different approaches and associated main references are given in Table 1.

Table 1. Different traditional approaches of Traffic Safety Evaluation (TSE)

Name	References
I. Before-after Observation	
• Naïve before-and-after study	(Council 2009; Ogden 1996)
• Before-and-after study with yoked comparison	(Griffin and Flowers 1997; Harwood et al. 2003; Ogden 1996)
• Before-and-after study with comparison group	(Council 2009; Sayed et al. 2006)
• Before-and-after study with the empirical bayes approach	(Council 2009; Elvik 2008; Hauer 1997; Namjune and Ihn 2013; Persaud and Lyon 2007)
II. Black Spot Identification Program	(Elvik et al. 2009; Hoque et al. 2006; Meuleners and Fraser 2008; Roper and Turner 2008)
III. Mainstream/Statistical Modeling Approaches	
• Descriptive model	(Archer 2005; OECD 1997; Rumar 1985)
• Predictive or analytical model	(Austin and Carson 2002; Chin and Quddus 2003; Datta et al. 2001; Eenink et al. 2008; Hall 2000; Hankey et al. 1999; Hsiao 1986; Jovanis and Chang 1986; Kumara and Chin 2003; Lord et al. 2007; Lord et al. 2005; McCullagh and Nelder 1989; Miaou 1994; Miaou and Lord 2003; Oh et al. 2006; Saccomanno and Buyco 1988; Stanton and Salmon 2009)
• Risk model	(Brehmer 1994; Michon 1989; Näätänen and Summala 1976; Shinar 1978; Van Der Molen and Boetticher 1987; Williams et al. 1995)
• Accident consequence model	(Hutchinson 1986; OECD 1997; Pham 2008; Walz et al. 1986)
IV. Road safety audit (RSA)/ inspection (RSI)	(Appleton 1996; Austroads 1994; Hoque 1997; iRAP 2012; Job 2012; Ogden 1996; Sabey 1993)
V. Safe system approach	(ATC 2011; Chen and Meuleners 2011; Mooren et al. 2011a; Mooren et al. 2011b; MoT 2010; Roper and Turner 2008; Tingvall 1998; WHO 2010b)

Despite its long history, this established technique has some inherent problems, mainly associated with the use of accident data and its reliance on personal judgement. Such shortcomings tend to be magnified in developing counties.

3 Assessing Need for Alternative Safety Evaluation Approaches

Currently used traditional approaches have been critically reviewed, with particular focus on their principles, strengths and weaknesses. Requirements for the successful application of those approaches and application challenges, particularly for developing countries, have been highlighted. A detailed review can be seen in the Mahmud et al. 2016. The primary dependent variable of the successful application of traditional approach is the historical accident record. This accident record or accident database is often questioned due to significant underreporting, biasness, inconsistent reporting, coding and locational errors. These problems are disproportionately higher in the developing countries, like Bangladesh. Underreporting, coupled with unavailability of reported accident data, is common in developing countries (Hoque and Mahmud 2009; TRL 2003). Different studies concluded that under-reporting of fatalities varies from around 2 to 5% in high income countries and 25 to 50% in low or middle income countries (Hoque and Mahmud 2009; WHO 2013, 2015).

The study reported here has made a comprehensive assessment of accident data quality in Bangladesh with a view to assess the need for an alternative safety evaluation approach. The study has considered Bangladesh as case study area because the country ranks amongst the highest in terms of road fatality and injury rates. Though, officially there are around 2,500 fatalities and 3,000 grievous and simple injuries on Bangladesh roads each year, the actual estimated fatalities are as high as from 12,000 to 20,000 per year (ARI 2015; Hoque and Mahmud 2009; WHO 2015). The safety problem is very severe by international standards with some around 60 to 150 fatalities per 10,000 motor vehicles in Bangladesh compared to around 25, 16, 2 and 1.4 in India, Sri Lanka, the USA and UK, respectively. In addition, Bangladesh has a reliable road crash database for a particular segment of highway section maintained by a separate authority. The Police reported database has also been used to evaluate the extent of underreporting. Moreover, the road safety problems in Bangladesh have received previous attention (Hoque and Mahmud 2009; Mahmud et al. 2009; Mahmud et al. 2014; Mahmud et al. 2016). Finally, the Accident Research Institute (ARI) in Bangladesh is dedicated to accident research and investigation. This institution continues to lead in terms of road safety research and practice, including crash database development and investigation, as well as the assessment of counter measures for developing countries.

3.1 Accident Data Quality in Bangladesh: Case Study

There are many sources for road accident related data including records from police, hospitals, insurance companies and newspaper reporting. Each of the sources has limitations, biasness and errors. Police are officially responsible for reporting and recording of road accidents and casualties in Bangladesh. Data from other sources, such as medical or insurance data is scare. No formalized system has been developed for recording and collecting accident data from other sources in Bangladesh. The Accident Research Institute (ARI), BUET is continuing its effort to develop newspaper based accident database but the newspapers have large reporting inconsistencies and highlight generally only major fatal accidents, particularly in the nearby core areas around growth centres. Injury accidents or remote areas' accidents are almost unseen in the newspaper reports. Newspaper reporting quality is not quite significant and elaborated with respect to fact finding analysis and research.

It is also recognized by road safety practitioners in Bangladesh, including the Bangladesh Road Transport Authority (BRTA) and the Police, that police reported road traffic accident database is neither complete nor entirely an accurate record of all road accidents. The degree and diversity of the reporting and recording inconsistency, as well as the key stages of misreporting of road traffic accidents and injuries in Bangladesh are summarized in the following sub-sections.

3.1.1 Underreporting of Accidents

Overall Underreporting: Underreporting of road traffic accidents and injuries, coupled with incomplete and inconsistence recording of information of reported accidents, are the major recognized drawbacks in relation to road safety analysis and evaluation based on historical data, particularly in Bangladesh.

According to the police reported official statistics, there were at least 1782 fatalities and 928 injuries in 1755 reported accidents in 2013 (ARI 2015). A study of Bangladesh Road Accident Costing conducted by the Transport Research Laboratory (TRL) in 2003 showed that there was 1,700,301 accidents; among them 11,049 was fatal, 102,136 were grievous, 311,890 minor and 1,275,226 were property damage accidents (Silcock 2003). This study also estimated that the total casualties resulting from road traffic accidents were 529,880, among them 12,792 fatalities and 165,464 seriously injured (TRL 2003). This estimation used a household survey. Comparing data with official police reports, the study suggests that the police record only one-third of all traffic fatalities and 2% of serious injuries. Another study, Bangladesh Health & Injury Survey Report, January 2005, estimated that the total number of fatalities due to traffic accidents is not less than 10,000 per year and for every injury death, 50 injured attend emergency care (i.e. 500,000 people are estimated to visit emergency room (Rahman 2005). World Health Organisation (WHO) estimated the number of deaths due to road accidents country by country. Table 2 provides a comparative summary of the WHO estimation and police reported death. The range of differences between WHO estimation and police reported death is 5 to 12.

3.2 Two Controlled Studies

Differences between ARF and FIR: The first-hand source accident data is the police reported First Information Report (FIR) (in Bengali 'Khatian') which is transferred to the Accident Report Form (ARF) at the thana (police station) level. The police at the thana records information filling the prescribed form ARF separately for each road accident.

Table 2. WHO estimation and police reported official statistics

WHO report	Year	Modelled number of road traffic deaths		Police reported accident deaths	Adjusted factor
		Point estimate	90% Confidence interval		
(WHO 2010a)	2007	20038	14882–29155	3749	5
(WHO 2013)	2010	17289	15415–19164	2443	7
(WHO 2015)	2013	21316	17349 – 25283	1782	12
Average		19548		2658	8

The ARF form is compiled in the police range and is input in the prescribed software called Microcomputer Accident Analysis Package (MAAP) by trained police personnel. From a pilot study conducted by ARI in some selected thanas, it was found that almost one third of the FIR reported accidents do not appear neither in the ARF nor in MAAP (Table 3). A significant number of accidents are being under-reported.

Table 3. Comparison between police station FIR and official record of accident data

Accident by severity	Fatal accident			Non-fatal accident		
Police station (PS)	PSFIR	Database	Missing	PS FIR	Database	Missing
Savar	160	90	44%	64	32	50%
Dhamrai	64	33	48%	43	14	67%
Saturia	21	13	38%	8	5	38%
Manikganj	37	25	32%	28	19	32%
Ghior	27	23	15%	17	17	0%
Shibalay	56	49	13%	29	30	−3%
TOTAL	365	233	36%	189	117	38%

Source: (ARI 2010)

Difference between BBA and Police Records: The Bangladesh Bridge Authority (BBA), evolved from the Jamuna multipurpose Bridge Authority (JMBA) and the maintenance operators of Jamuna Bridge and approaches. The BBA maintain a separate accident database on a regular basis solely for this roadway segment.

Initial analysis revealed that the data reported by BBA is more reliable in terms of number and the depth of reporting, consistency and accuracy (Hoque et al. 2008). Table 4 provides a comparative summary between BBA and police reported data. The extent of underreporting of police database is evident. Police report on average 9 and 11.6 times less than the BBA, in terms of accidents and casualties, respectively.

Table 4. BBA and Police record of accident

Year	Accidents			Casualties		
	BBA	Police	Adjustment factor	BBA	Police	Adjustment factor
2011	156	20	7.8	481	49	9.82
2012	144	24	6	433	50	8.66
2013	172	8	21.5	395	14	28.2
Total	472	52	9.08	1309	113	11.6

Source: Bangladesh Police and BBA

Figure 1 compares the police reported accidents with BBA in the periods 2004 to 2006 and 2011 to 2013. In 2006, police reports were 41% of the BBA and it was almost double than the previous year. On the other hand, in 2013, police reports were only 5% compared to BBA and it is less than half of the previous year. It is clear that the number of underreporting, as well as the quality of the police accident database is deteriorating with the time.

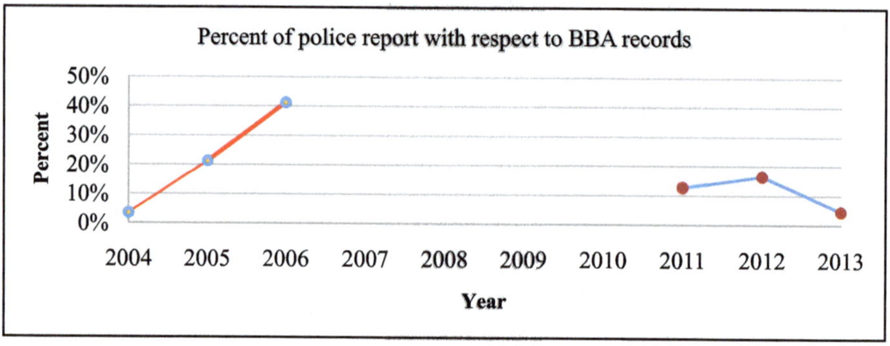

Fig. 1. Percent of police report with respect to BBA record

3.2.1 Injury and PDO Accidents: A Neglecting Issue

It is reported that, for every death by road traffic accidents, around 6 to 150 injuries are occurring depending on the level of reporting and definition of accidents related injuries (Table 5). In developed countries, this ratio is more than 50 and most of the middle income countries and a number of developing countries, such as Srilanka, Myanmar and Bhutan, it is between 15 and 5 (WHO 2009) (Table 5). In Bangladesh, according to TRL 2003, "the number of casualties being hospitalised is at least 53 times that reported and best estimated at 73 times and minor injuries have an even larger adjustment factors, as they are the most common injury type and the least likely to be reported to the police". According to UNICEF, 2005 "For every injury death, 50 injured attend emergency care and 500,000 people are estimated to visit emergency room" (Rahman 2005). According to BBA records, on the Jamuna bridge (JMB) and Approach Road, number of injuries is more than 8.2 times higher than the deaths and non-fatal accidents is 3.67 times higher than the fatal accidents (Table 6).

Table 5. Injuries and death ratios in different countries

Country	Fatalities	Injuries	Injuries/Death	Source of data
Japan	6639	1034445	156	Police data
Germany	4949	431419	87	Federal Statistics Office data
United Kingdom	3298	264288	80	Police data
USA	42442	3305237	78	Health data
Austria	691	53211	77	Statistics Austria
Canada	2889	199337	69	Police data
Belgium	1067	65850	62	Statistics Belgium estimates
Sweden	471	26636	57	Police data
Hungary	1232	27452	22	Police data
Australia	1616	31204	19	Health and Transport data
Srilanka	2334	31688	14	Police data 2007
Brazil	35155	407685	12	State Traffic Departments,
Ethiopia	2517	24792	10	Police data
Myanmar	1638	12358	8	Police and Transport data
Bhutan	111	724	7	Police data
Saudi Arabia	6358	36025	6	Police data

Source: (WHO 2009)

Table 6. Accident and Casualties according to BBA on JMB

Year	Fatal Accident	Accidents	Fatalities	Casualties	Fatalities/Accident	Non-fatal/fatal accident	Injuries/Death
2011	34	156	55	481	0.35	3.59	7.7
2012	31	144	44	433	0.31	3.65	8.8
2013	36	172	43	395	0.25	3.78	8.2
Total	101	472	142	1309	0.30	3.67	8.2

Source: Bangladesh Bridge Authority (BBA)

On the other hand, Bangladesh police mainly handle those accidents in which one or more persons are killed. They do not monitor accidents involving injured persons. As a result, most of the injured accidents are not reported by the police. Even in reported fatal accidents, the numbers of injuries are not reported adequately or are simple ignored. Tables 7 and 8 provide recent five years' police reported accidents and casualties' statistics with ratio between fatal and non-fatal accidents and casualties, respectively. From the Table 7, it is seen that non-fatal accidents are less than one third of fatal accidents. Moreover, the average number of reported injury is around 65% of reported deaths.

Due to lack of continued training and monitoring, the level of under-reporting has increased, particularly injury and property damage only (PDO) accidents, which occur on the newly constructed road network in remote areas. The quality of the reported data also has deteriorated.

Table 7. Accidents statistics in Bangladesh

Year	Fatal	Total Accident	Non-fatal accident	Grievous./ Fatal	Simple/Fatal	PDO/Fata	Non-fatal/Fatal
2009	2161	2815	654	0.22	0.03	0.05	0.30
2010	1911	2437	526	0.20	0.03	0.04	0.28
2011	1566	2017	451	0.20	0.05	0.04	0.29
2012	1515	1939	424	0.19	0.06	0.04	0.28
2013	1421	1755	334	0.16	0.05	0.02	0.24
Average	1715	2193	478	0.19	0.05	0.04	0.28

Source: Bangladesh Police and Accident Research Institute (ARI)

Table 8. Casualties statistics in Bangladesh

Year	Deaths	Total Casualties	Grievous Inj./Deaths	Simple Inj./Deaths	Injury/Death
2009	2703	4449	0.53	0.11	0.65
2010	2443	4149	0.52	0.18	0.70
2011	2072	3520	0.52	0.18	0.70
2012	1953	3295	0.44	0.25	0.69
2013	1782	2710	0.35	0.47	0.52
Average	2191	3625	0.47	0.24	0.65

Source: Bangladesh Police and Accident Research Institute (ARI)

3.2.2 Inconsistent and Incomplete Reporting

Inconsistency and incompleteness in the reported records are also major problems. In spite of having 67 parameters in the ARF, many of the attributes remain empty or incomplete. Significant inconsistences are also found among the reported attributes. Inconsistency between description of accident and vehicle type, type of collision, collision diagram, movement of vehicle, movement or position is common in the report form, as well as electronic database.

Driver and victim demography: Attributes related to demography of the victim including age, pattern of injury, location of victim during accident and action during accidents, is missing in many report forms. More than 40% of fatal and 50% of injured persons' age is not reported in the database (Table 9). The attributes related to injury

Table 9. Age Missing

Year	All casualties		All casualties known age		Percent of missing	
	Fatalities	Injuries	Fatalities	Injuries	Deaths	Injuries
2001	2388	2565	1663	1505	30%	41%
2002	3053	3285	1831	1446	40%	56%
2003	3334	3740	2039	1666	39%	55%
2004	3150	3026	1732	1246	45%	59%
2005	2960	2570	1540	1051	48%	59%

Source: Bangladesh Police and Accident Research Institute (ARI)

pattern remain blank in around 90% of cases. Location, such as front, rear, rooftop or hanging and action such as sitting, standing, alighting or boarding during accidents, are missing in around 50% to 80% of cases.

Location of accident: Missing location of accident, which is needed for different type of traditional approach based studies, is one of the single biggest problems with the quality of reported data. Whereas, the exact location of accident is important for any investigation or to select appropriate treatments for any hazardous sites. Identification of black spots or hazardous locations is mainly based on the recurrence of accidents in a particular location. Problems related to identification of accident locations from the reported records or lack of exact location, has been acknowledged for a long time as a fundamental deficiency in accident records. Such incomplete data can thus jeopardize the success of safety analysis.

In Bangladesh, accident location is provided based on road inventory. There is no modern system for accident location identification. The latest version of chain age inventory is almost 20 years old and that is only available for national highways. Due to changing of road alignment or developing by-pass road of many national highways, that inventory is now mostly out of date. Therefore, in spite of having three stage location identifications systems in the report form, locational attributes are simply missing. There is no location identification system for feeder, local or city road except some parts of Dhaka, the capital city. An exhaustive effort has been made by a team of researchers to identify the reported national highway accident locations using all of the available means, including complete report form, road inventory and detailed road map. Around 50% of accident locations were identified (Table 10).

Table 10. Accident on major National highways in Bangladesh (1998–2009)

Route	Reported accidents	Approximate location sorted out	Unidentified
N 1 (Dhaka-Cox-bazar)	2600	1720	34%
N 2 (Dhaka-Sylhet)	1644	1079	34%
N 3 (Dhaka-Mymensing)	797	242	70%
N 4 (Dhaka-Jamalpur)	992	371	63%
N 5 (Dhaka- Dinajpur)	2809	1496	47%
N 6 (Bogra-Rajshahi)	772	552	28%
N 7 (Faridpur – Patuakhali)	1272	831	35%
N 8 (Dhaka-Mawa)	692	288	58%
Average unidentified			46%

Source: Bangladesh Police and Accident Research Institute (ARI)

Accident causes: The cause of road accidents is one of the most important parameters in any safety analysis. In the accident report, there are three options to provide the factors of a particular accident, namely primary, secondary and tertiary factors.

However, causes of accidents are not reported properly in almost 90% cases in the police reported accident statistics. According to that report, 98% accidents are happening primarily due to careless driving or over speeding. There is no vehicle and road factor (Fig. 2). Whereas, according to BBA reports, around 12% vehicle factor and 4% road factor for the newly build national highway accidents. According to PIARC, 3% of accidents are happening solely due to road factors and 2% due to vehicle factors. In consideration of combined factors, the share of road and vehicle is 34 and 13%, respectively (Mikulik and Hollo 2007). Secondary and tertiary factors are totally missing in most of the police reported Accident Report Form (ARF). Therefore, it is very difficult to evaluate the exact reasons for accidents using this database.

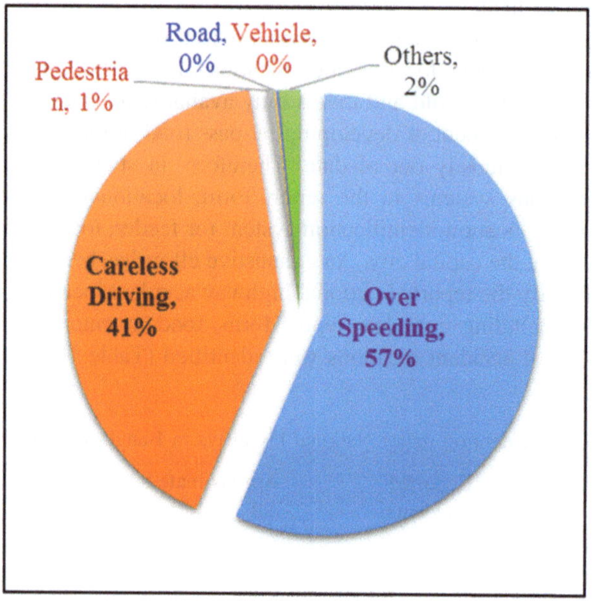

Fig. 2. Primary factors of accident according to police report

From the above analysis, it is clear that the reported accident data is highly underreported and there is lack of consistent detailed information in the reported records in Bangladesh, as well as most of the developing countries. Many important parameters are missing which have significant influence in analysis and modelling approach for evaluation. Apart from this, accumulation of sufficiently large number of sample data for analysis to obtain a statistically sound result is time consuming. Longer collection cycle to gather sufficient data for analysis is a major problem in accident based safety evaluation due to low frequency of accidents. Therefore, it is debatable whether these approaches are always ethical, as they demand sufficiently large number of accidents and eventually deaths, injuries and property losses before any intervention.

There are some other experimental challenges with these traditional approaches related to casual effects, such as treatment, exposure, trend and randomness. Sometimes these effects cannot be easily explained through modelling (Council 2009; LGED 2015; Ogden 1996). In addition, there are some other specific issues with each individual method of this general approach (Archer 2005; OECD 1997; Yang 2012).

Therefore, there is a need for alternative approaches, particularly for developing countries, that will produce more effective and reliable safety measures without the need for (or in addition to) accident data.

4 Alternative Approaches for Safety Evaluation

Many ways of commissioning non-accident data have been developed as surrogate approaches to overcome the heavy reliance on historical accident data. Some reliable approaches are speed variances, traffic violence, erratic manoeuvres and traffic conflicts (Debnath et al. 2014). Among these surrogate measures, using traffic conflict techniques (TCTs) for diagnosing road safety problems has gained acceptance as a proactive surrogate measure (Sayed et al. 2013; Zaki et al. 2013; Zheng et al. 2014). Several studies have been conducted for the development, validation of traffic conflict techniques using different measures based on mainly speed, time, distance, acceleration and/or deceleration, as a substitute to the historical accident records in traffic safety research. A detailed review of the development and application of TTCs can be seen in the (Zheng et al. 2014). Almost all the past research has been based on lane based car dominated homogeneous traffic conditions found mainly in developed countries. The traffic environment in developing countries is very different, where non-lane based mixed traffic operation is common. Lane discipline and level of enforcement for guiding driver behaviour in right way is very poor. The presence of non-motorized vehicles and large number of pedestrians, not only in urban roads but also in rural highways, has induced the complexity of traffic operation with wide variety of characteristics (Khan and Maini 1999).

However, the concept and theory behind the traffic conflict techniques could easily be calibrated and applied for the developing countries traffic environment. Evasive action based subjective observation and judgment approaches can easily be applied in any traffic situations of developing countries e.g. (Muhlrad 1993). Objective approaches, in which certain parameters are measured and applied in different simple equations as indicators, is the most valid and reliable technique for traffic conflict evaluation. Earlier, there was a major challenge regarding the accurate measurement of parameters, such as absolute and relative speed, distance, acceleration, deceleration and potential conflict point, particularly in the complex traffic environment of developing countries. Recently, with the development of advanced image processing systems, video analysis techniques have been introduced to address this challenge. Through these techniques, conflict data can be extracted automatically from the video sensors. Data collection using video sensors is now a widely advocated procedure, which addresses many issues related to manual data collection through providing a more efficient and reliable way to collect, store and analyse traffic behaviour data. Moreover, video data offers a permanent, verifiable account of road user behaviour. Furthermore, the precise estimation of exposure, as well as other parameters of road safety analysis can be greatly improved by analysing the position of a road user in space and time, i.e. vehicle trajectories (Ismail et al. 2009; Saunier et al. 2010).

More recently, traffic simulators are being used for observing traffic behaviour and safety evaluation, through evaluating different parameters and applying in different traffic conflict measures (Vlahogianni 2013). Moreover, microscopic traffic safety simulation modelling is also a significant advancement for improving reliability of conflict study and application the technique in different traffic environment (Young et al. 2014). Integration of those advancements may offer a useful perspective in traffic conflict studies of different traffic environments, thus providing a sustainable way for understanding the most concerning traffic safety problems.

5 Conclusions

Sustainability in traffic safety evaluation is crucial for optimum use of resources through identifying problem oriented appropriate preventive measures, as well as for achieving the goals of traffic safety improvement. The practicing traditional approaches have a long history, in terms of their robustness, methodological development, research and application. However, these approaches are mainly depending on the reported accident records which are often questioned due to some major limitations, especially in developing countries, like Bangladesh. This paper provided a detailed assessment on the quality of accident data with a case study of Bangladesh. Some key aspects are:

- Underreporting and unavailability of accident and related information is common (25 to 50% in developing countries). In Bangladesh, the range of differences between WHO estimation and police reported death is 5 to 12. In compare to BBA reported data, police reports are, on an average, 9 and 12 times less than the BBA reports, in terms of accidents and casualties respectively.
- Almost one third of reported accidents to the police station are not reported in the accident database.
- Non-fatal accidents and injuries are significantly under-registered. According to police reports, injuries are less than fatalities, whereas, according to statistics, for every death by road traffic accidents, around 6 to 150 injuries are occurring.
- Reported records are significantly inconsistent and incomplete. More than 40% fatal and 50% injured persons' age is not reported in the reported database. The attributes related to injury pattern remain blank in around 90% form.
- Locational attributes are simply missing (more than 50% of national highway accidents). There is no location identification system for feeder, local or city roads, except for some parts of Dhaka city in Bangladesh.
- Causes of accidents are not reported properly in almost 90% of cases in the police records in Bangladesh.
- There are some other issues related to accident data such as codding error, biasness, longer collection cycles which often restrict the ability to identify the problem characteristics and to evaluate the causative factors for selecting appropriate interventions.

- Finally, in Bangladesh, the quality of police reported accident database is deteriorating, particularly in recent past years.

Accident data is highly under-reported and lacks of consistency in detailed information in Bangladesh records, as well as those of most of developing countries. Therefore, there is a need for alternative approaches for developing countries that will produce more effective and reliable safety measures without the need for (or in addition to) accident data. A framework for using traffic conflict techniques (TCTs) in developing counties has been suggested here. It is expected that incorporation of advanced image processing system and microscopic traffic simulation modelling approach for defining conflict can provide an easy, wider and sustainable platform for safety evaluation in developing countries.

Traffic and accident characteristics in developing countries are very different from those in developed countries. To address those issues, there is a need for some specific research for a more accurate representation of driver behaviour to measure conflict and evaluate safety situation. Some priority research areas include:

- define standard threshold values for different indicators in different traffic environments, including non-lane base heterogeneous traffic environment;
- develop new measures for identifying conflict or safety problems to address some special types of single vehicle accident (e.g. overturn induced by pothole; shoulder drops & bridge approach drops, tyre burst induced by over loading and over use of tyres, passenger falling from the rooftop& freight top);
- undertake more controlled studies in non-lane base heterogeneous rural traffic environments;
- develop relationships between traffic conflicts and crashes for different traffic environments;
- use of different surrogate safety indicators for conflict and severity analysis;
- explore the modelling of overtaking conflicts; and place more emphasis on the analysis of non-lane base heterogeneous traffic environments; and
- develop traffic safety micro-simulation models with traffic safety indicators for developing counties.

References

Appleton, I.: Progress with the introduction of road safety audit in Australia and New Zealand. Paper presented at the 18th ARRB Transport Research Conference & Transit New Zealand Land Transport Symposium (1996)

Archer, J.: Indicators for traffic safety assessment and prediction and their application in micro-simulation modelling: A study of urban and suburban intersections. Ph.D. thesis, Royal Institute of Technology, Stockholm, Sweden (2005)

ARI. Road Accident in Bangladesh: Problem of Data Analysis. Retrieved from Accident Research Institute (ARI) BUET, Bangladesh (Internation report, unpublished) (2010)

ARI. Road Safety Facts in Bangladesh (Facts). Retrieved from Accident Research Institute (ARI), BUET, Bangladesh (2015). http://www.buet.ac.bd/ari/Downloads.php

ATC. National Road Safety Strategy 2011 – 2020. Retrieved from Australian Transport Council (ATC) (2011)

Austin, R.D., Carson, J.L.: An alternative accident prediction model for highway-rail interfaces. Accid. Anal. Prev. **34**(1), 31–42 (2002)

Austroads. Road safety audits. Publication AP-30/94. Austroads, Sydney (1994)

Brehmer, B.: Psychological aspects of traffic safety. Eur. J. Oper. Res. **75**(3), 540–552 (1994)

Chen, H., Meuleners, L.: A literature review of road safety strategies and the safe system approach (2011)

Chin, H.C., Quddus, M.A.: Applying the random effect negative binomial model to examine traffic accident occurrence at signalized intersections. Accid. Anal. Prev. **35**(2), 253–259 (2003)

Council, T.S.: Before-and-After Study Technical Brief. Institute of Transportation Engineers (2009)

Datta, T., et al.: Using GIS to analyze statewide traffic crash data in Michigan. Paper presented at the Proceedings of The Aet European Transport Conference, Held 10–12 September, 2001, Homerton College, Cambridge, Uk-CD-Rom (2001)

Debnath, A.K., et al.: Proactive safety assessment in roadwork zones: a synthesis of surrogate measures of safety. Paper presented at the Proceedings of the 2014 Occupational Safety in Transport Conference (2014)

Eenink, R., et al.: Accident prediction models and road safety impact assessment: recommendations for using these tools. Institute for Road Safety Research, Leidschendam (2008)

Elvik, R.: The predictive validity of empirical Bayes estimates of road safety. Accid. Anal. Prev. **40**(6), 1964–1969 (2008)

Elvik, R., et al.: The handbook of road safety measures: Emerald Group Publishing (2009)

Griffin, L.I., Flowers, R.J.: A discussion of six procedures for evaluating highway safety projects: The Division (1997)

Hall, D.B.: Zero-inflated Poisson and binomial regression with random effects: a case study. Biometrics **56**(4), 1030–1039 (2000)

Hankey, J., et al.: Identification and evaluation of driver errors: Task C report, driver error taxonomy development. Project no. Dtfh-61-97-c-00051. Virginia Tech Transportation Institute, Blacksburg, VA (1999)

Harwood, D., et al.: Safety effectiveness of intersection left-and right-turn lanes. Transp. Res. Rec. J. Transp. Res. Board **1840**, 131–139 (2003)

Hauer, E.: Observational Before/After Studies in Road Safety. Estimating the Effect of Highway and Traffic Engineering Measures on Road Safety (1997)

Hoque, M.M.: Road Safety Audit in Developing Countries. Retrieved from transportation Research Group, Department of Civil and Environment Engineering, University of Southampton, UK (1997)

Hoque, M.M., Mahmud, S.S.: Road Safety Engineering Challenges in Bangladesh. Accident Research Institute. Bangladesh university of Engineering and Technology (2009)

Hoque, M.M., et al.: Road Safety Hazards At Jamuna Multipurpose Bridge (JMB) Site: Implications for Bridge Management. Paper presented at the 23rd ARRB Conference–Research Partnering with Practitioners, Adelaide, Australia (2008)

Hoque, M.M., et al.: Observational studies of hazardous road locations on national highways in Bangladesh. Paper presented at the ARRB Conference, 22nd, 2006, Canberra, ACT, Australia (2006)

Hsiao, C.: Analysis of Panel Data, Econometric Society Monograph No. 11. Cambridge University Press, Cambridge (1986)

Hutchinson, T.P.: Statistical modelling of injury severity, with special reference to driver and front seat passenger in single-vehicle crashes. Accid. Anal. Prev. **18**(2), 157–167 (1986) iRAP. Vaccines for Roads, 2nd edn. Retrieved from International Road Assessment Program (iRAP), Hampshire, UK (2012)

Ismail, K., et al.: Automated analysis of pedestrian-vehicle conflicts using video data. Transp. Res. Rec. J. Transp. Res. Board **2140**, 44–54 (2009)

Job, R.S.: Advantages and disadvantages of reactive (black spot) and proactive (road rating) approaches to road safety engineering treatments: When should each be used? Paper presented at the Australasian Road Safety Research Policing Education Conference, 2012, Wellington, New Zealand (2012)

Jovanis, P.P., Chang, H.-L.: Modeling the relationship of accidents to miles traveled. Transp. Res. Rec. **1068**, 42–51 (1986)

Khan, S., Maini, P.: Modeling heterogeneous traffic flow. Transp. Res. Rec. J. Transp. Res. Board **1678**, 234–241 (1999)

Kumara, S., Chin, H.C.: Modeling accident occurrence at signalized tee intersections with special emphasis on excess zeros. Traffic Inj. Prev. **4**(1), 53–57 (2003)

LGED. Rural Road Safety Manual for LGED. Retrieved from Local Government Engineering Department (LGED), Bangladesh, June 2015

Lord, D., et al.: Further notes on the application of zero-inflated models in highway safety. Accid. Anal. Prev. **39**(1), 53–57 (2007)

Lord, D., et al.: Poisson, Poisson-gamma and zero-inflated regression models of motor vehicle crashes: balancing statistical fit and theory. Accid. Anal. Prev. **37**(1), 35–46 (2005)

Lund-University. The Swedish Traffic Conflict Technique. Retrieved from Sweden (2014)

Mahmud, et al.: Road Safety Problems in Bangladesh: Achievable Target and Tangible Sustainable Actions. Jurnal Teknologi, 70(4) (2014)

Mahmud, et al.: Traditional Approaches to Traffic Safety Evaluation (TSE): Application Challenges and Future Directions. Bridging the East and West, 242 (2016)

Mahmud, et al.: Road Safety Problems in Bangladesh: Some Major Initiatives, Constraints and Requirements. Transport and Communications Bulletin for Asia and the Pacific, 61 (2009)

McCullagh, P., Nelder, J.A.: Generalized Linear Models, vol. 37. CRC Press (1989)

Meuleners, L., Fraser, M.: Review of the Wa state black spot program: a literature review of Australian and international black spot programs. Centre For Population Health Research, School of Public Health, Curtin University of Technology, Perth. [Links] (2008)

Miaou, S.-P.: The relationship between truck accidents and geometric design of road sections: Poisson versus negative binomial regressions. Accid. Anal. Prev. **26**(4), 471–482 (1994)

Miaou, S.-P., Lord, D.: Modeling traffic crash-flow relationships for intersections: dispersion parameter, functional form, and Bayes versus empirical Bayes methods. Transp. Res. Rec. J. Transp. Res. Board **1840**, 31–40 (2003)

Michon, J.A.: Explanatory pitfalls and rule-based driver models. Accid. Anal. Prev. **21**(4), 341–353 (1989)

Mikulik, J., Hollo, P.: Road Accident Investigation Guidelines for Road Engineers. World Road Association PIRAC Technical Committee (2007)

Mooren, L., et al.: Safe System–International Comparisons of this Approach. Paper presented at the A Safe System-making it happen: Proceedings of the Australasian College of road Safety Conference, Melbourne (2011a)

Mooren, L., et al.: Safe system–comparisons of this approach in Australia. Paper presented at the Australasian College of Road Safety National Conference, Melbourne (2011b)

MoT. Safe Journeys: New Zealand's Road Safety Strategy 2010-2020. Retrieved from Wellington: Minsitry of Transport (MoT), New Zealand (2010)

Muhlrad, N. Traffic conflict techniques and other forms of behavioural analysis: Application to safety diagnoses: na (1993)

Näätänen, R., Summala, H.: Road-user behaviour and traffic accidents. Publication of: North-Holland Publishing Company (1976)
Namjune, P., Ihn, L.Y.: Comparison of Empirical Bayes Method and Before-After Study Method. Paper presented at the Proceedings of the Eastern Asia Society for Transportation Studies (2013)
OECD. Road safety principles and models: review of descriptive, predictive, risk and accident consequence models (1997)
Ogden, K.W.: Safer roads: a guide to road safety engineering. Institute of Transport Studies, Department of Civil Engineering, Monash University, Melbourne, Australia (1996)
Oh, J., et al.: Accident prediction model for railway-highway interfaces. Accid. Anal. Prev. **38**(2), 346–356 (2006)
Persaud, B., Lyon, C.: Empirical Bayes before–after safety studies: lessons learned from two decades of experience and future directions. Accid. Anal. Prev. **39**(3), 546–555 (2007)
Pham, M.H., Mouzon, O., Chung, E., El Faouzi, N.E.: Sensitivity of road safety indicators in normal and crash cases. Paper presented at the 10th International Conference on Application of Advanced Technologies in Transportation, Athens, Greece (2008)
Rahman, A.: Bangladesh health and injury survey: report on children: Directorate General of Health Services Ministry of Health an (2005)
Roper, P., Turner, B.: Why do we need to take a risk assessment based approach in road safety? Paper presented at the ARRB Conference, 23rd, 2008, Adelaide, South Australia, Australia (2008)
Rumar, K.: The role of perceptual and cognitive filters in observed behavior. In: Human Behavior and Traffic Safety, pp. 151–170. Springer (1985)
Sabey, B.: Safety audit procedures and practice. Paper presented at the Ptrc Traffex 93 Conference Proceedings. Seminar On Traffic Management And Road Safety, Tuesday 20 April 1993, National Exhibition Centre, Birmingham (1993)
Saccomanno, F.F., Buyco, C.: Generalized loglinear models of truck accident rates (1988)
Saunier, N., et al.: Large-scale automated analysis of vehicle interactions and collisions. Transp. Res. Rec. J. Transp. Res. Board **2147**, 42–50 (2010)
Sayed, T., et al.: Safety evaluation of stop sign in-fill program. Transp. Res. Rec. J. Transp. Res. Board **1953**, 201–210 (2006)
Sayed, T., et al.: Automated safety diagnosis of vehicle–bicycle interactions using computer vision analysis. Saf. Sci. **59**, 163–172 (2013)
Shinar, D.: Psychology on the road. The human factor in traffic safety (1978)
Silcock, R.: Guidelines for estimating the cost of road crashes in developing countries. London, Department for International Development Project, 7780 (2003)
St-Aubin, P.G.: Traffic Safety Analysis for Urban Highway Ramps and Lane-Change Bans Using Accident Data and Video-Based Surrogate Safety Measures. Paper presented at the Masters Abstracts International (2012)
Stanton, N.A., Salmon, P.M.: Human error taxonomies applied to driving: A generic driver error taxonomy and its implications for intelligent transport systems. Saf. Sci. **47**(2), 227–237 (2009)
Tingvall, C.: The Swedish'Vision Zero'and how Parliamentary approval was obtained. Paper presented at the Road Safety Research, Policing, Education Conference, 1998, Wellington, New Zealand, vol. 1 (1998)
TRL. Bangladesh Road Crash Costing Discussion Document Retrieved from Roads and Highways Department (RHD), Bangladesh (2003)
Van Der Molen, H., Boetticher, A.: Risk models for traffic participants: a concerted effort for theoretical operationalizations. road users and traffic safety (1987)

Vlahogianni, E.I.: Modeling duration of overtaking in two lane highways. Transp. Res. Part F Traffic Psychol. Behav. **20**, 135–146 (2013)

Walz, F., et al.: The car-pedestrian collision, injury reduction, accident reconstruction, mathematical and experimental simulation, head injuries in two wheeler collisions. Interdisciplinary Working Group for Accident Mechanics, University of Zurich and Swiss Federal Institute of Technology (1986)

WHO. Global status report on road safety: time for action: World Health Organization (2009)

WHO. Data systems: a road safety manual for decision-makers and practitioners (2010a)

WHO. Global Plan for the Decade of Action for Road Safety 2011-2020 Retrieved from World Health Organization (WHO), Geneva (2010b)

WHO. WHO global status report on road safety 2013: supporting a decade of action: World Health Organization (2013)

WHO. Global status report on road safety 2015: World Health Organization (2015)

Williams, A.F., et al.: Factors that drivers say motivate safe driving practices. J. Saf. Res. **26**(2), 119–124 (1995)

Yang, H.: Simulation-based evaluation of traffic safety performance using surrogate safety measures. Rutgers University-Graduate School-New Brunswick (2012)

Young, W., et al.: Simulation of safety: A review of the state of the art in road safety simulation modelling. Accid. Anal. Prev. **66**, 89–103 (2014)

Zaki, M., et al.: Application of computer vision to diagnosis of pedestrian safety issues. Transp. Res. Rec. J. Transp. Res. Board **2393**, 75–84 (2013)

Zheng, L., et al.: Traffic conflict techniques for road safety analysis: open questions and some insights. Can. J. Civil Eng. **41**(7), 633–641 (2014). doi:10.1139/cjce-2013-0558

Evaluation of Railroad Ballast Field Degradation Using an Image Analysis Approach

Maziar Moaveni[1], Erol Tutumluer[1(✉)], John M. Hart[2], and Mike McHenry[3]

[1] Department of Civil and Environmental Engineering, University of Illinois at Urbana-Champaign, Urbana, IL, USA
{moavenil, tutumlue}@illinois.edu

[2] Computer Vision and Robotic Laboratory, University of Illinois at Urbana-Champaign, Urbana, IL, USA
jmhart3@illinois.edu

[3] Transportation Technology Center, Inc., Pueblo, CO, USA
mike_mchenry@aar.com

Abstract. Identifying the level of ballast degradation generally involves ballast sampling and mechanical sieve analyses in the laboratory, which can be time consuming, laborious and costly. As an automated alternative, image processing techniques has the potential to directly and objectively assess ballast condition and degradation levels from high resolution images of ballast layers captured in the field or reproduced in the laboratory. This paper presents the development stages and implementation of an innovative image processing method for assessing the degradation levels of ballast using ballast cross section images collected in the field and also reproduced in the lab. Advanced image enhancement methods, including gamma adjustment, histogram equalization, and bi-lateral image filtering, combined with image segmentation techniques such as watershed algorithm and image thresholding, were used to successfully extract size and shape properties of individual ballast particles as a mean to quantify the level of ballast degradation. In order to capture images of the ballast layers in the field, a detailed procedure was developed to ensure the resulting images captured would perform consistently and accurately when processed with the machine vision algorithms. Rapid imaging of a large quantity of ballast samples was needed for producing ground truth data to be used as input into the machine vision algorithms. The results of this study showed that the images captured in the field and the images captured in the lab from the corresponding collected ballast samples looked quite different. This confirmed that a robust image processing algorithm which can be linked to indices based on sieve analysis methods needs to be adjusted/trained from the images and samples collected in the field. The findings of this ballast field and lab imaging study showed promising future potential of the described image processing technique for replacing the tedious and time consuming ballast sampling and sieve analysis processes for evaluating ballast degradation.

J. Pombo and G. Jing (eds.), *Recent Developments in Railway Track and Transportation Engineering*, Sustainable Civil Infrastructures, DOI 10.1007/978-3-319-61627-8_8

1 Introduction

Railroad ballast is uniformly-graded coarse aggregate placed between and immediately underneath the crossties to provide drainage as well as structural support for the repeated loading from train traffic. Ballast degradation is generally referred to change in particle size distribution of the ballast layer as the result of particle breakage and abrasion, migration of fine-grained soil from the subgrade or introducing coal dust from overloaded freight cars. Aggregate breakdown and crushing accounts for nearly 80% of the ballast degradation cases encountered in the field (Selig and Waters 1994). Maintaining an existing track with satisfactory serviceability requires periodic evaluation of the ballast condition with accumulated freight tonnage, and corresponding solution of the recurring ballast problems in problematic areas. Increased levels of degradation may lead to ballast differential settlement as well as alteration of the track geometry (Chiang 1989; Han and Selig 1997; Ebrahimi et al. 2012).

Two indices are commonly used in the North American railroad system to quantify ballast degradation conditions. These are (i) Fouling Index (FI) and (ii) Percentage Fouling (PF) (*1*). FI is the summation of percentage by weight of ballast material passing No. 4 (4.75 mm) sieve and the percentage passing No. 200 (0.075 mm) sieve. PF is the ratio of the dry weight of the material passing 3/8 in. (9.5 mm) sieve to the dry weight of the total sample. To estimate the constantly changing amounts of larger aggregate particles and the accumulated fine materials, traditional methods for degradation assessment involve on-site visual inspection by experts, as well as ballast sampling and testing via sieve analyses in the laboratory. Obviously, visual inspections are highly dependent on the level of track inspector's experience and hence very subjective in nature. To estimate field FI values for the underlying ballast layer, several samples of ballast need to be collected at different depths. However, ballast sampling as a labor intensive and time consuming process has its own difficulties, i.e., on how to collect truly representative samples when only limited quantities are allowed due to the concerns of changes in the geometry of the track. Moreover, the estimated FI values could vary drastically, i.e. field variability, depending on the locations along the track where ballast samples are collected. Like any sampling method, this method may miss the worst field conditions. To achieve a more reliable evaluation of the functioning of the existing ballast layer for various track conditions and thus enhance serviceability of the track, it is imperative that aggregate size and shape properties of ballast be objectively monitored in a consistent and preferably automated fashion along the track for quantifying the ballast degradation trends throughout its service life.

Machine Vision (MV) technology recovers useful information regarding a scene projected onto a camera sensor to form a two-dimensional image. This technology has been used in many different areas of science and engineering to develop inspection, gauging and counting systems (Jain et al. 1995). Thus, machine-vision-based inspection systems, once trained on example images, have the potential for directly assessing ballast condition and estimating the FI values from images of ballast layers captured in the field. Novel imaging systems that capture shape and size properties of individual railroad ballast particles have been reported in previous research studies (Descantes et al. 2006; Boler et al. 2012; Moaveni et al. 2013, 2016). Digital image and Discrete

Element Modeling (DEM) methodology has been developed to study effects of aggregate particle size and morphological characteristics on ballast performances (Huang 2009). Later, a modified approach for transforming ballast particle size distribution to constriction size distribution was proposed by implementing a three dimensional (3D) imaging method (Sun et al. 2014). Imaging systems incorporate advanced image processing techniques such as segmentation and 3D reconstruction. The results of these techniques, in terms of classifying ballast sources according to their shape properties as well as capturing the change in morphological properties under degradation process, are promising; however, they still require ballast sampling and individual particle scanning before applying the image processing algorithms. It is therefore highly desired to have an automatic inspection system that is capable of evaluating the condition of the in-service ballast layer without the need for individual particle scanning. With a machine vision based inspection system in place, ballast field deterioration levels can be continuously monitored over various time periods at different locations along the track, such as in the shoulder or crib, and particularly in special track work and transition zones. Additionally, spatial variation of fouling levels versus the ballast depth profiles associated with the use of different clean ballast types or other geographic and loading considerations can also be recorded. It is envisioned that, once such technology is developed, it can be installed on Shoulder Ballast Cleaners (SBC) with minimal effort in research and development to collect ballast data during maintenance activities on shoulders. Further, the information gathered from such an inspection system can then be incorporated into a railroad company's ballast maintenance database. Combined with Geographical Information System (GIS) output, it should provide the necessary foundation for a reliable and robust assessment of the needed for a comprehensive ballast management tool.

2 Objective and Scope

The primary objective of this paper is to describe the initial development of a machine vision based inspection approach for in-service ballast condition assessment and to demonstrate the capability of such a system in assessing ballast aggregate properties and/or degradation zones from high resolution images captured in the field. A comprehensive procedure was established as part of this research to capture high resolution trench images from in-service ballast cross sections under ambient field conditions. Then, a family of image processing techniques including bi-lateral filtering, gamma adjustment and watershed segmentation were integrated into an innovative machine vision algorithm. A new imaging based index called "Percent Degraded Segments" or PDS which can be linked to Selig's FI values was developed as part of this study. This imaging based approach is envisioned to ultimately establish an automated visual inspection system that is capable of evaluating the condition of in-service ballast layer without the need for field sampling and laboratory testing. The performance of the developed algorithm has been tested and validated to investigate the effect of different types of crossties on the extent of ballast degradation.

3 Description of Ballast Image Acquisition in the Field

To collect field trench images of in-service ballast cross sections, the Facility for Accelerated Service Testing (FAST) for Heavy Axle Load (HAL) applications at Transportation Technology Center (TTC) in Pueblo, Colorado was selected. In total 4 zones in Sect. 3 of High Tonnage Loop (HTL) with four different types of concrete crossties and spacing with or without under-tie-pads were nominated for this task because all of these zones had experienced similar levels of train traffic. After selecting the desired locations, one crosstie was removed and vertical trenches with a target depth of 3 ft. (91-cm) perpendicular to the track were dug with a backhoe to expose the cut section of in-service ballast layers in the field between the two adjacent crossties. Efforts were taken to minimize the disturbance of the ballast walls so that an accurate image of the ballast layers beneath the ties could be taken (see Fig. 1).

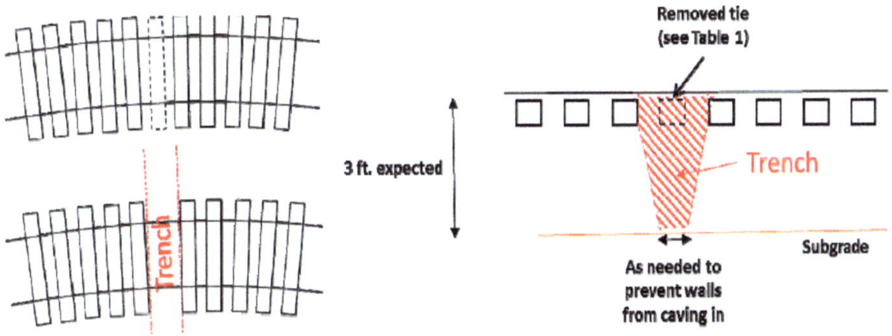

Fig. 1. Plan and cross-sectional views of typical trenches dug in Sect. 3 HTL.

Table 1 summarizes the four zones in terms of crosstie numbers corresponding to the location of trenches as well as the ballast properties and types and arrangement of the concrete crossties. Rubber under-tie-pads and/or "half frame" ties were used to decrease the stress intensity at the interface of ballast layer and the crosstie. Therefore, it is generally expected that less degradation would be observed in the ballast layer under these conditions. More information related to the influence of types of concrete crossties and under-tie-pads on the ballast life can be found elsewhere (Akhtar et al. 2012; McHenry et al. 2015). Two major objectives were perused during the field imaging effort. First, identifying the performance of the machine vision inspection system developed as part of this study. Second, to investigate the effect of different types of crossties on severity of ballast degradation. As part of scheduled research activities, trenches were made laterally across the track so the condition of the ballast within and underneath the various tie zones could be visually observed (see Table 1).

A 15.1-megapixel DSLR camera was installed on a tripod in an inverted configuration and a T-Bar extension was used to penetrate the space inside the trench. The camera was set to aperture priority mode, to allow settings to reduce distortion in the images. Additionally, a 1-in. (25.4 mm) diameter white ball was positioned in the field

Table 1. Crosstie information for the trench images analyzed

Trench ID (crosstie no.)	Zone no.	Type of concrete crosstie	Under tie pads	Crosstie spacing (in.)
1100	1	Half-Frame	No	24
1177	1	Half-Frame	No	24
1274	2	Conventional	No	24
1388	3	Conventional	Yes	24
1477	4	Conventional	Yes	24

of view of the camera at all times for the purpose of calibrating the spatial resolution captured in the images. The images were captured from seven areas inside each trench to give a comprehensive coverage of the face of the ballast cross section underneath the crossties. Figure 2 illustrates the final field image acquisition set up as well as the designation of areas of interest for imaging underneath the crossties. In total, 35 high resolution images were captured and stored for processing. A visual ranking number from 1 (low level of degradation) to 5 (high level of degradation) was manually assigned to each category of the images from identical imaging locations but at different trenches. This way, a rough estimation of degradation levels at various locations underneath various types and arrangements of crossties was identified.

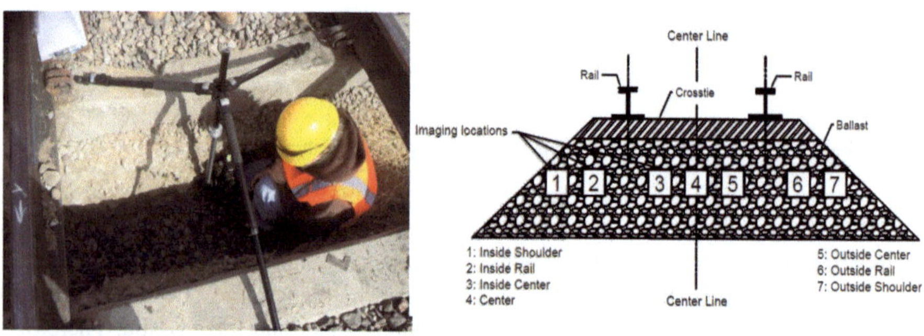

Fig. 2. Final imaging setup (left) and ballast cross section imaging locations (right).

The main considerations in capturing the images involved proper exposure, consistent spatial resolution and a reduction of any distortions of the ballast particles in the image. Adequate sunlight was cast evenly across the section for proper exposure but needed to be diffused in order to not cast shadows, in addition, a triple exposure setting was used to capture three consecutive images; the auto metered exposure and one image with a setting lighter and another darker exposure. This ensured that one or more field images would be suitable for processing since these images could not be retaken once the area was refilled with ballast for continuing train operations. To maintain a consistent spatial resolution, number of pixels used per unit length, the optical axis of the camera lens was made normal to the cross section and the aperture setting was adjusted to F8.0 to provide the least lens distortion. In addition, the camera zoom was adjusted such that the field of view covered the same surface area (24 in. by 16 in.) in each image (providing approximately 80 pixels/cm). This ratio was verified by determining the amount of pixels covering a 1-in. (25.4 mm) diameter calibration ball also placed in the ballast image view. This step was essential to determine the particle measurements in each image and was consistently adopted during image acquisition to ensure the uniformity of the output data from the machine vision algorithm, described later in detail.

4 Description of Ballast Image Processing Procedure

The flowchart shown in Fig. 3 illustrates the general procedure of the developed image based inspection system. A core component of this algorithm is a three-stage image processing module, which includes a pre-processing step, a segmentation step and a post-processing step. Given a ballast image, the algorithm starts with a suitable image enhancement technique, usually gamma correction, amplifying the visual differences between the bright or dark pixels in the image (Jobson et al. 1997; Gonzalez and Woods 2001). In some cases when the ballast image appears foggy or overexposed, where the majority of pixels of the image are concentrated in a short band of intensities, a histogram equalization technique (Gonzalez and Woods 2001; Pisano et al. 1998) can be used to enhance the image contrast. The resulting image is then converted to a grayscale image. Next, bilateral image filtering is applied to this grayscale image to reduce the surface texture. Bilateral filtering (Paris et al. 2008) is a technique primarily aimed at suppressing those internal variations while retaining and at times also visually enhancing the boundaries between objects. Bilateral filtering is controlled by two

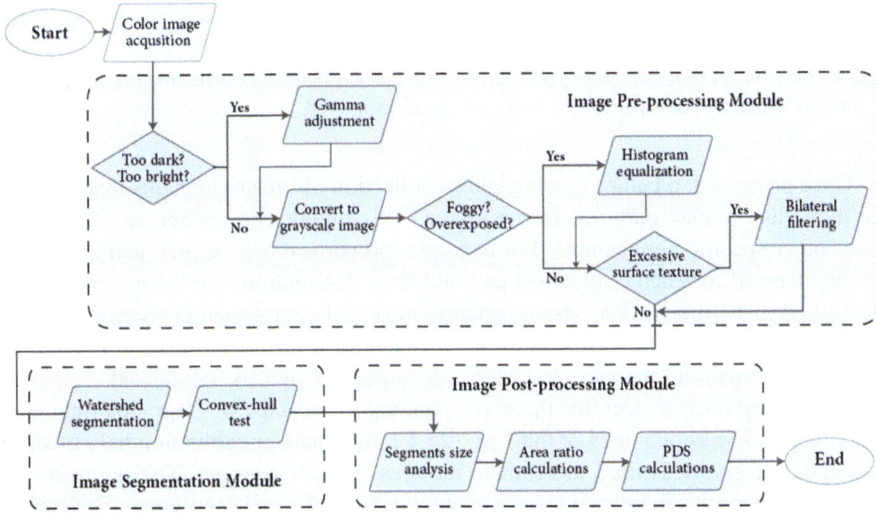

Fig. 3. Flowchart of the algorithm for the imaging-based ballast inspection system.

parameters including spatial Gaussian width and range Gaussian width. Effects of these parameters on the filtered image can be found elsewhere (Paris et al. 2008). This concludes the pre-processing step. The output of the preprocessing step is the input of the segmentation step, which is done via a fine-tuned watershed segmentation algorithm. The watershed algorithm has been shown to perform better than other leading algorithms in the presence of mutually touching particles, such as ballast (Vincent and Soille 1991). In addition, the performance of the algorithm is improved whenever internal texture details are suppressed as discussed earlier. When individual rock analysis is desired, the segments are passed through a convex hull test to filter out true and false segments.

Individual ballast particles are generally expected to be convex or at least "nearly convex." This hypothesis is checked on each of the segments that are identified during the watershed segmentation step. The convexity test constructs for each segment its convex hull and computes the ratio between the area (number of pixels) of the segment and the area of the convex hull. A segment is rejected (i.e., declared not to be an actual ballast particle) if the above ratio falls below a certain adjustable threshold. Such a threshold value in the range of 0.6 to 0.75 is usually preferable (see Fig. 4).

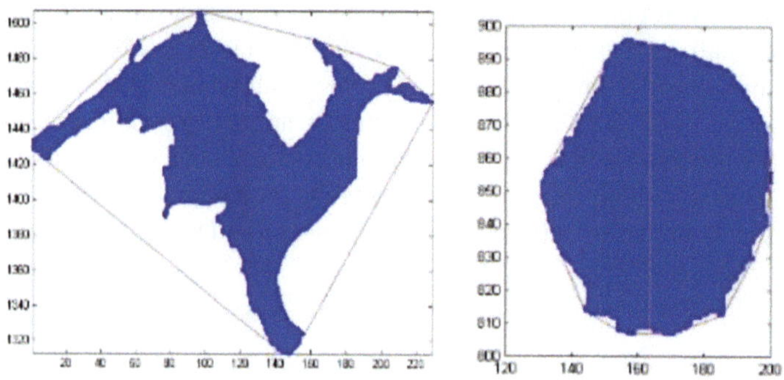

Fig. 4. (a) Convex hull (boundary) of a false-segment (non-ballast particle) and (b) Convex hull of a true-segment (ballast particle) – [axes are pixel locations]

Once processed, a ballast image yields a collection of image segments that estimate the individual ballast particles in the image. This output can further be utilized for many other specific applications. For example, properties such as size and angularity can be assessed for each output segment and their distributions can be estimated over the entire ballast image. The visual description of different image processing steps is depicted in Fig. 5 which shows an example image of a ballast cross section.

The degradation analysis on each segmented image is conducted using an area-based approach to identify the corresponding Percentage of Degraded Segments (PDS) value. The known area of the 1-in. (25.4 mm) diameter calibration ball, in terms of number of pixels, is used to estimate the area of each segment. The segments are then partitioned into three size classes including typical, small and large. The typical category represents average size ballast particles, the small category represents severely

Fig. 5. Visual description of image processing steps for an example image of a ballast cross section.

degraded particles, and the large represents oversized areas with particles too small to be identified individually, such as fine-grained soils. These categories are determined by normalizing the areas with respect to the area of the calibration ball and setting thresholds at less than 60% for small and more than 300% for large. The latter categories are labelled as Degraded Segments.

A score for each image, i.e. PDS, is defined as the percentage of the total area of large and small segments compared to the total area of the ballast image. Let s_i (i = 1, …, n) be the areas of the segments *1*, …, *n* in the image. Let *b* be the area of the 1-in. (25.4-mm) diameter calibration ball and *A* be the area of the image. Note that all areas are measured in terms of number of pixels. Then, the PDS value can be computed following Eq. 1:

$$Define : J = \{1 \leq i \leq n | threshold_s < \frac{s_i}{b} < threshold_l, threshold_s = 0.6, threshold_l = 3\}$$

$$Then, PDS(\%) = 100 \times (1 - \frac{\sum i \in J s_i}{A}) \quad (1)$$

Note that achieving desirable segmentation of each ballast image at a certain degradation level needs interaction between user and the algorithm during the image processing task through a trial and error type procedure. This includes fine-tuning the segmentation parameters related to bilateral filtering and water segmentation in an effort to decrease the segmentation error. During the analyses of ballast images, it was observed that the performance of the segmentation algorithm in terms of precise detection of ballast particles and/or degradation zones was dependent on the level of particle size variability in each image. In other words, identifying one single set of segmentation parameters capable of detecting both small and large ballast particles for the entire 24 in. by 16 in. image was found to be very challenging. Consequently, it was decided that each ballast image would initially be cropped into three identical sub-images representing top, middle and low portions of the captured ballast cross section. The reason for creating three sub-images is also explained by the increasing level of ballast degradation with ballast layer depth, which was clearly indicated in previous studies (Moaveni et al. 2013). Therefore, the variability of particle sizes processed in each sub-image is expected to be less than the variability of particle sizes if the entire image is considered. Through visual observation of the segmentation results, this approach proved to increase the performance of the algorithm. Thus, three sets of segmentation parameters and PDS values were recorded for each ballast image. The final PDS value for each image was reported as the average of three PDS values associated with three sub-images. The main segmentation parameters and their description used in the proposed image processing algorithm are described briefly in this section.

4.1 Spatial Gaussian Width and Range Gaussian Width

The bilateral filtering of an image at pixel position p relative to pixel position q with intensity I is given in Eq. 2 as follows:

$$BF[I_p] = \frac{1}{w_p} \sum\nolimits_{q \in s} G_{\sigma_s}(\|p - q\|)\, G_{\sigma_r}(|I_p - I_q|)\, I_q \qquad (2)$$

where $\frac{1}{w_p}$ is the normalization factor, and G_{σ_s} and G_{σ_r} are the Gaussian spatial kernel and the Gaussian range kernel. Equation 2 is a normalized weighted average where G_{σ_s} is a spatial Gaussian that decreases the influence of distant pixels, G_{σ_r} a range Gaussian that decreases the influence of pixels q with an intensity value different from I_p. Consider the interested pixel at position p, the Spatial Gaussian width, represented by σ_s, determines the size (number of pixels) of the Gaussian window used to smooth images. In other words, Spatial Gaussian width is the size of the considered pixel neighborhood (thinner than $2\sigma_s$) used to calculate the average intensity of pixel at position p (Chaudhury and Sage 2001). Also, because bilateral filtering retains the boundaries of ballast particles, one needs the Range Gaussian width, represented by σ_r, to represent the minimum amplitude of an edge, so that the pixels in the considered neighborhood whose intensity is above or below this boundary are ignored in the smoothing process. In ballast images with 3158 × 4152 pixels, σ_s and σ_r could range from 3 to 200, but since some ballast boundaries before the preprocessing stage are already blurred, it is preferable to keep below 20 and below 10.

4.2 Strel-Size

The strel size describes the radius (number of pixels) of a disk-shaped morphological structuring element used to clean up the ballast particles thus enhancing the segmentation results. It is found that in ballast images captured in the laboratory with 3158 × 4152 pixels, the strel-size needs to be 12 to 16 to segment 1-in. (25.4 mm) ballast particles, 16 to 20 for 2-in. (50.8 mm) particles, and 20 to 30 for 3-in. (76.2 mm) particles.

4.3 Number of Pixels Within the Boundary of the Calibration Ball

The number of pixels for the 1-in. (25.4 mm) calibration ball means how many pixels are needed to represent 1-in. (25.4 mm) diameter in the image. The user can obtain this value at the pre-processing stage.

4.4 Convex-Hull Threshold

The convex hull of a set X of points in the Euclidean space is the smallest convex set that contains X. For instance, when X is a bounded subset of the plane, the convex hull may be visualized as the shape enclosed by a rubber band stretched around X (de Berg et al. 2000). In the developed algorithm, the convex-hull threshold falls between 0.6 and 0.75.

5 Segmentation Results and Discussion

To evaluate the performance of the developed machine vision based ballast inspection algorithm, initially a set of subjective visual inspection results were developed using a scale from 1 to 5, 1 being the best for clean ballast and 5 being the worst for heavily degraded ballast. These visual rankings and the PDS values for all the 35 collected images of in-service ballast layers are summarized in Tables 2, 3, 4, 5, 6, 7 and 8. Since human inspectors are mostly capable of comparing ballast images and ranking them by degradation levels from 1 to 5, the visual rankings were then compared with the results produced by the PDS scores. For the majority of the images, the two rankings were found to be compatible. In particular, it was observed that in both rankings, ballast degradation levels were higher at locations closer to the center of the track as opposed to shoulder and outside rail areas. This phenomenon is likely to be related to load distribution patterns and stress concentration effect at the central areas of ballast layer.

Table 2. Ballast layer visual rankings and percentages of degraded segments at inside shoulder imaging location (see Fig. 2)

Ballast cross section image		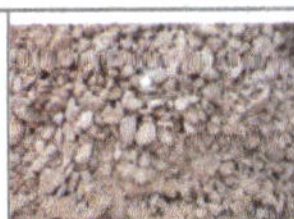			
Trench ID	1477	1177	1274	1100	1388
Visual degradation ranking	1	2	3	4	5
PDS (%)	48.9	52.5	50	65.4	65.4

Table 3. Ballast layer visual rankings and percentages of degraded segments at inside rail imaging location (see Fig. 2)

Ballast cross section image				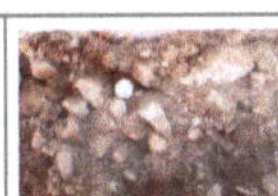	
Trench ID	1177	1100	1477	1388	1274
Visual degradation ranking	1	2	3	4	5
PDS (%)	45	73.4	73.9	93.1	63.54

Table 4. Ballast layer visual rankings and percentages of degraded segments at inside center imaging location (see Fig. 2)

Ballast cross section image					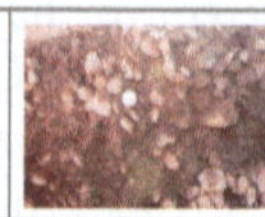
Trench ID	1177	1100	1388	1274	1477
Visual degradation ranking	1	2	3	4	5
PDS (%)	74.4	63.1	77.94	86.4	87.84

Table 5. Ballast layer visual rankings and percentages of degraded segments at center imaging location (see Fig. 2)

Ballast cross section image					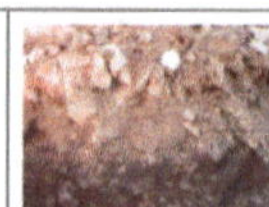
Trench ID	1177	1100	1388	1477	1274
Visual degradation ranking	1	2	3	4	5
PDS (%)	75.94	42.97	90.27	83.24	89.87

Note that exposing the entire cross section of ballast layer underneath the tie is a destructive and not always a feasible alternative to evaluate the condition of substructure for an in-service track. Nevertheless, ballast condition at the shoulder location can still be considered as a good representation of the overall condition of ballast cross section. This is true especially for comparing degradation levels at different locations along the track.

Both the visual rankings and the image segmentation results confirmed that the trench images 1100 and 1177 from various imaging locations underneath the crossties had the lowest severity of degradation. After perusing the types and arrangements of concrete ties, it was found that these two trenches were actually located underneath the half frame concrete ties. This finding could be expected since half frame crossties are designed to decrease the stress levels on the ballast layer by increasing the contact area between the tie and the top surface of ballast.

Note that in some cases the visual rankings and the PDS-based rankings did not entirely match in the same sense that is explained above. One explanation for this finding is the impact of the image acquisition environment, in particular the light conditions, on the segmentation results. Under or overexposure as well as shadows distort segmentation, usually leading to a higher fouling rate than the correct ones (see Table 8 for the ballast image at 1100 image location as an example of overexposure). Another reason is that some of the ballast images are visually similar in their fouling levels, making the human ranking a subjective one. A third reason is that selection of

Table 6. Ballast layer visual rankings and percentages of degraded segments at outside center imaging location (see Fig. 2)

Ballast cross section image		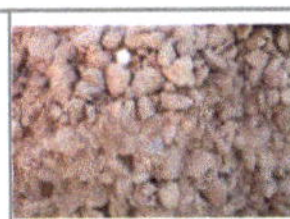	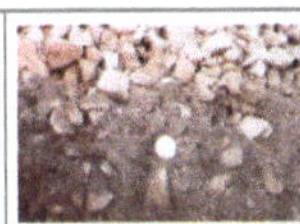	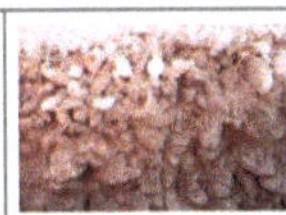	
Trench ID	1177	1100	1274	1477	1388
Visual degradation ranking	1	2	3	4	5
PDS (%)	45.40	53.5	79.76	81.54	79.45

Table 7. Ballast layer visual rankings and percentages of degraded segments at outside rail imaging location (see Fig. 2)

Ballast cross section image					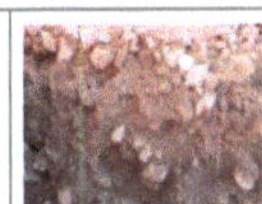
Trench ID	1177	1100	1274	1388	1477
Visual degradation ranking	1	2	3	4	5
PDS (%)	58.37	63	67.13	75	77

Table 8. Ballast layer visual rankings and percentages of degraded segments at outside shoulder imaging location (see Fig. 2)

Ballast cross section image					
Trench ID	1100	1177	1477	1274	1388
Visual degradation ranking	1	2	3	4	5
PDS (%)	62.1	55.94	66.57	66.8	85.17

segmentation parameters such as convex hull threshold for different images may also introduce differences in the accuracy of the estimated PDS. A PDS value calculated from the segmentation results with optimal convex hull threshold is more accurate than one that is calculated with a suboptimal convex hull threshold.

Based on the discussion above, it is important to point out that even though segmentation results and visual ranking did not match with each other completely, the ballast images that showed clear signs of significant degradation resulted in higher PDS values while the ballast images that showed little sign of degradation resulted in lower PDS scores. As the next step, the current and ongoing work has been directed at improvement of the developed image analysis and segmentation technique by making use of ground truth ballast samples collected from the trench walls containing the ballast captured in the images taken, processing them in the laboratory for FI values and then linking them to imaging based PDS values from the analyses of the ballast cross section images.

6 Conclusions and Recommendation

This study presented the preliminary results related to an ongoing research project which is mainly focused on the development of machine vision based inspection systems for an objective assessment of in-service ballast degradation in the field. A combination of image processing techniques including image enhancement through gamma adjustment and an innovative bi-lateral filtering and watershed segmentation was developed. The post-processing analysis on the segmentation results included utilizing the convex hull concept on the black and white binary images to differentiate between the zones that are close to the shape of ballast particles from those that relate to the spaces between the particles.

The developed algorithm was used to assess the conditions of 35 images of in-service ballast layers captured in the field. The initial findings of field implementation showed that the described image processing technique could successfully classify these in-service ballast sections in terms of their level of degradation. More or less a close match was observed between the image segmentation results and the visual ranking orders that was assigned to each ballast image. The image processing methods used in this research could also very well highlight the improved performance of half frame crossties in decreasing the level of ballast degradation.

Further development and improvement of the ballast field imaging technology can eventually be quite helpful as an inspection tool when Ballast Shoulder Cleaners (BSC) can be installed with cameras to automate the collection of ballast images from the cut shoulders below the crossties during operation. With this vision, SBCs equipped with this technology would become a continuous data collection device for automatic ballast inspection. In addition, this technology advancement could not only be used to provide degradation values for the entire length the SBC traverses, but also be used to better make recommendations of follow-up undercutting procedures to remove degraded ballast along specific areas.

The current focus of this research effort is also the development of an Imaging Based Fouling Index (IBFI) which can be correlated to the commonly used Selig's FI values as ground truth using ballast samples both in the laboratory and from the field captured images. The routine application of the validated technology innovation will establish an IBFI database from an automated ballast inspection system. Implementation of this machine vision based technology in an automated fashion can be considered as an essential element of a comprehensive Ballast Management System (BMS). Additionally, it can eliminate the need for spot checking and ballast sampling to ultimately provide an effective on-site ballast evaluation and management tool for assisting practitioners in sustainable and timely selection of the proper maintenance and rehabilitation strategies.

Acknowledgements. The Association of American Railroads (AAR) and TRB Innovations Deserving Exploratory Analysis (IDEA) program have supported the preliminary development of the described machine vision algorithm field imaging of ballast through the research project collaborations of Rail Transportation and Engineering Center (RailTEC) at the University of Illinois and the Transportation Technology Research Inc. (TTCI). The authors of this paper highly appreciate the contributions and technical support from Dave Davis, senior scientist, and Dr. Dingqing Li, Chief of FRA programs, at TTCI. The contents of this paper reflect the views of the authors who are responsible for the facts and the accuracy of the data presented herein. This paper does not constitute a standard, specification, or regulation.

References

Selig, E.T., Waters, J.M.: Track Geotechnology and Substructure Management. Thomas Telford Publications, London (1994)

Chiang, C.C.: Effects of water and fines on ballast performance in box tests. Master of Science Degree Project Report No. AAR89-366P, University of Massachusetts (1989)

Han, X., Selig, E.T.: Effects of fouling on ballast settlement. In: Proceedings of the 6th International Heavy Haul Railway Conference, Cape Town, South Africa (1997)

Ebrahimi, A., Tinjum, J.M., Edil, T.B.: Protocol for testing railway ballast in large-scale cyclic triaxial equipment. Geotech. Test. J. **35**(5), 762–804 (2012)

Jain, R., Kasturi, R., Schunck, B.G.: Machine Vision. McGraw-Hill Inc, New York (1995)

Descantes, Y., Fosse, Y., Milcent, F.: Automated measurement of railway ballast angularity. J. Mater. Civ. Eng. **18**(4), 612–618 (2006)

Boler, H., Wnek, M., Tutumluer, E.: Establishing linkages between ballast degradation and imaging based aggregate particle shape, texture, and angularity. In: Proceedings of the 2nd International Conference on Transportation Geotechnics, Hokkaido, Japan (2012)

Moaveni, M., Wang, S., Hart, J.M., Tutumluer, E., Ahuja, N.: Evaluation of aggregate size and shape by means of segmentation techniques and aggregate image processing algorithms. Transp. Res. Rec. J. Transp. Res. Board, 2335, 50–59 (2013)

Moaveni, M., Qian, Y., Qamhia, I.I.A., Tutumluer, E., Basye, C., Li, D.: Morphological characterization of railroad ballast degradation trends in the field and laboratory. Transp. Res. Rec. J. Transp. Res. Board, 2545, 89–99 (2016)

Huang, H.: Discrete element modeling of railroad ballast using imaging based aggregate morphology characterization. Ph.D. Dissertation. University of Illinois, Urbana, pp. 1–3 (2009)

Sun, Y., Indraratna, B., Nimbalkar, S.: Three-dimensional characterization of particle size and shape for ballast. Géotechnique Letters **4**(3), 197–202 (2014)

Akhtar, M., Davis, D., LoPresti, J.: Increasing strength and reducing lifecycle cost of concrete tie track. In: Railway Track and Structures, pp. 16–19 (2012)

McHenry, M., LoPresti, J.: Concrete Tie Testing at FAST to Address Improved Track Strength. Technology Digest TD-15-013. AAR/TTCI, Pueblo, Colorado (2015)

Jobson, D., Rahman, Z., Woodell, G.: A multiscale retinex for bridging the gap between color images and the human observation of scenes. IEEE Trans. Image Process. **6**(7), 965–976 (1997)

Gonzalez, R., Woods, R.: Digital Image Processing, 2nd edn. Prentice Hall, Upper Saddle River (2001)

Pisano, E., Zong, S., Hemminger, M., Deluca, M., Johnston, R., Muller, K., Braeuning, M., Pizer, S.: Contrast limited adaptive histogram equalization image processing to improve the detection of simulated spiculations in dense mammograms. J. Digit. Imaging **11**(4), 193–200 (1998)

Paris, S., Kornprobst, P., Tumblin, J., Durand, F.: Bilateral filtering: theory and applications. Found. Trends Comput. Graph. Vis. **4**(1), 1–73 (2008)

Vincent, L., Soille, P.: Watersheds in digital spaces: an efficient algorithm based on immersion simulation. IEEE Trans. Pattern Anal. Mach. Intell. **13**(6), 583–598 (1991)

Chaudhury, K.N., Sage, D., Unser, M.: Fast bilateral filtering using trigonometric range kernels, IEEE Trans. Image Process. **20**(11) (2001)

de Berg, M., van Kreveld, M., Overmars, M., Schwarzkopf, O.: Computational Geometry: Algorithms and Applications, pp. 2–8. Springer, Heidelberg (2000)

Influence of Subgrade Differential Settlement on Riding Performance of High-Speed Train

Yanmei Cao[1(✉)] and Jiting Qu[2]

[1] School of Civil Engineering,
Beijing Jiaotong University, Beijing 100044, China
ymcao@bjtu.edu.cn
[2] Faculty of Infrastructure Engineering, School of Civil Engineering,
Dalian University of Technology, Dalian 116024, China
qjt@dlut.edu.cn

Abstract. The differential settlement of subgrade may accelerate degradation of tracks, lower passenger comfort and increase derailment risk of trains, especially in high-speed railway. In this paper, a dynamic analysis model for train-ballasted track-subgrade system is established to study the effect of different settlement profiles in a deeper insight, in which the subgrade settlement is simulated by Gaussian function profiles with different amplitudes and longitudinal lengths. In the case study, the dynamic responses of train vehicles and track structure subjected to 100 kinds of differential settlements of subgrade are calculated with the CRH2 high-speed train at different speeds. Results show that the performance indicators of running trains increase obviously with the train speed; the accelerations of the passenger car-bodies increase not only with the amplitude of the settlement, but also depend on the length of the track deformation; the shorter is the longitudinal length of settlement, the bigger the derailment risk is; the greater is the amplitudes of subgrade differential settlement, the bigger the probability of poor passenger comfort and risk of derailment are.

1 Introduction

The ballasted track, as a traditional track structure, has been widely used in high speed railways due to good elasticity, low cost, easy maintenance, and obvious absorption to noise. However, the differential settlement of subgrade is a troublesome to designers and infrastructure mangers (Kang 2016). It may accelerate degradation of tracks, lower passenger comfort and increase derailment risk of trains, especially in high-speed railway.

Despite the past efforts to understand and mitigate the influence of track deformation on train-track interaction system, most studies have focused on the effect of short-wave track defects (Popo et al. 1999; Knothe and Grassie 1993; Dahlberg 2001; Wang 2007; Liu and Zhang 2009; Xu et al. 2012), rather than on the differential settlement of subgrade on the running safety and passenger comfort of trains. Issues related to track deformation due to the poor performance of the earthworks are difficult to detect and complex to solve because the subgrade and the subsoil are generally inaccessible. Abadi et al. (2016) described some current empirical ballast settlement

J. Pombo and G. Jing (eds.), *Recent Developments in Railway Track and Transportation Engineering*, Sustainable Civil Infrastructures, DOI 10.1007/978-3-319-61627-8_9

models, and evaluated them using experimental data generated using the Southampton Railway Testing Facility. Due to the limitation of empirical model, the model test and the numerical methods have been applied. Zou designed a 1:1 model test to study the influence of differential subgrade settlement on the ballasted track (Zou et al. 2011). Tatsuya et al. (2014) proposed an analytical procedure with iterative calculation by linear finite element analysis to estimate the cyclic plastic deformation of railroad ballast under repeated moving-wheel loads. Paixão used a FEM method to simulate different scenarios in normal railway lines and gave a parametric study (Paixão et al. 2015). However, the researches on the influence of differential subgrade settlement on the riding performance of ballasted high-speed railways are few.

Thus, this paper focuses on providing greater insight into the effect of subgrade differential settlement on the riding performance of high-speed railways. Herein, the finite/infinite element train-ballasted track-subsoil numerical model is established, considering the CRH2 high-speed train of China. The subgrade settlement is simulated by Gaussian function profiles with different amplitudes and longitudinal lengths. Also, the dynamic responses of train vehicles and track structure subjected to 100 kinds of differential settlements of subgrade are calculated, and how different subgrade settlement may affect track degradation, passenger comfort and running safety are analyzed.

2 Modelling of Train-Track-Subsoil Interaction System

A 2-dimensional interaction model of train-track-subsoil system is established by the software of ABAQUS, which consists of train, track, ballasted subgrade, and subsoil.

2.1 Train Model

The CRH2 high speed train of China is adopted, which can be seen Fig. 1 (left). The train marshaling is composed of eight vehicles (4 Motors + 4 Trailers), and each vehicle can be simplified as a multi-degree of freedom model with primary and secondary spring-damp suspension system whose stiffness and damping are denoted by k_1, c_1, k_2, c_2, respectively shown in Fig. 1 (right).

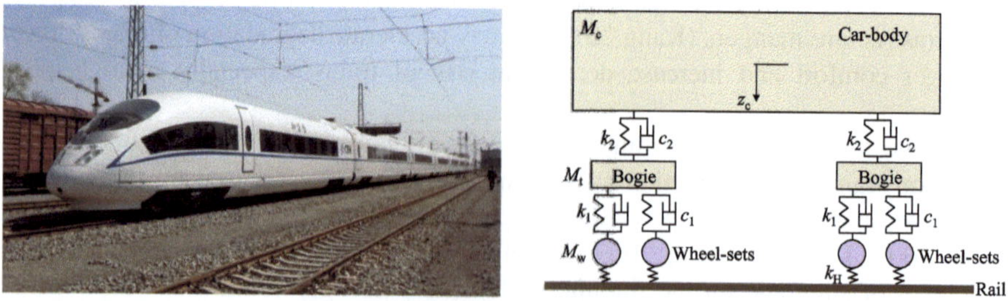

Fig. 1. CRH2 high speed train of China and the simplified model of a vehicle

In the modelling, only the vertical vibrations are taken into account. The car-body, the bogies and the wheelsets are simulated by solid elements and their weights are applied to the preassigned reference points by the form of point mass. The primary and secondary suspensions are simulated by linear point-to-point contact elements, while the vertical wheel-track contact is simulated by defining elastic surface-to-surface contact elements with the vertical stiffness of k_H and penalty function (Mijar et al. 2000). Besides, the transfer of force and the coordination of degrees of freedom are accomplished by multipoint-constraint elements. The main parameters of a CRH2 train are listed in Table 1.

Table 1. Parameters of CRH2 trains

Car body	Height/m	3.7
	Length/m	25
	Height of gravity centre/m	1.8
	Weight M_c/t	47.2
Bogie	Central distance $2s$/m	17.5
	Wheel-axle distance $2d$/m	2.5
	Height of gravity centre/m	0.51
	Weight M_t/t	3.2
Wheelsets	Radius/m	0.43
	Weight M_w/t	2
Primary suspension	Vertical stiffness k_1/(kN/m)	1000
	Vertical damping c_1/(kN•s/m)	18
Secondary suspension	Vertical stiffness k_2/(kN/m)	200
	Vertical damping c_2/(kN•s/m)	10
Wheel-rail	Contact stiffness k_H/(kN/m)	1.5×10^6

2.2 Track-Subgrade-Soil Model

The sketch of track-subgrade-soil interaction is shown in Fig. 2, in which the rail irregularity is not taken into account. The total length of the numerical model is 600 m so as to simulate the whole process of train incoming and leaving from the studied areas. The displacements of the bottom nodes of the model were restrained in the vertical direction; however, for both sides of the substructures below the sleepers, the infinite elements are adopted to avoid the reflection of vibration waves on the artificial boundary.

The use of infinite domain for the substructure boundary is based on the research work by Lysmer and Kuhlemeyer (1969), who proposed the first absorbing boundary conditions and use viscous boundary tractions (dashpots) to absorb incident waves. Also, Lysmer & Kuhlemeyer's viscous boundary is the only absorbing boundary available in ABAQUS, and it is meshed in a scanning way to form quadrangle plain strain infinite elements (CINPS4). Moreover, an implicit viscous boundary is unconditionally stable and the accuracy of the method is acceptable for engineering purposes. However, it is still advisable to leave a relatively large margin between the boundary and the central region of the model, which is the other reason of the model size of 600 m in this paper.

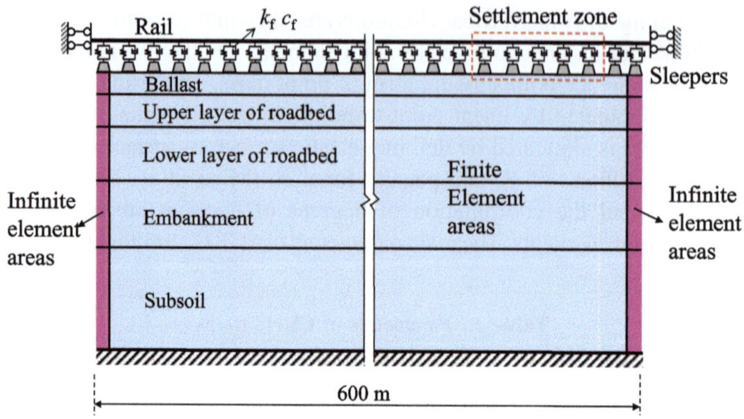

Fig. 2. Sketch of modelled track and substructures

The rails (HST60) are regarded as 2-dimensional continuous beam with the second moment of area of $2 \times 3.217 \times 10^{-5}$ (m^4), which is supported by the discrete sleepers. The element size is 0.15 m, and both the longitudinal displacement and the angular displacement are restrained at two ends of the rail model. The sleepers are simulated by solid elements and the distance between two adjacent sleepers is 0.6 m. The fasteners between rail and sleepers are simplified as linear point-point connectors with the coefficients of $k_f = 60$ MN/m and $c_f = 75$ kN•s/m. In addition, the subgrade and substructures are regarded to be elastic and simulated by the 4-node plane stress elements of CPSR4, and the element size is 0.15 m, and some parameters of the train-track-subgrade-soil model can be seen in Table 2.

Table 2. Parameters of track and substructures

Structures	Depth/m	Density/(kg/m^3)	Elastic modulus/MPa	Poisson ratio
Rail	–	7830	2.1×10^{11}	0.3
Sleeper	–	2100	3.6×10^4	0.2
Ballasted subgrade	0.35	1900	120	0.27
Upper layer of roadbed	0.7	2184	120	0.3
Bottom layer of roadbed	2.3	1939	70	0.3
Embankment soil	3.5	1837	50	0.35
Subsoil	30	1898	124	0.3

2.3 Simulation of Subgrade Differential Settlement

Usually, the cosine curve is adopted to simulate the differential settlement of ballasted subgrade (Dahlberg 2001; Xu et al. 2012; Zou et al. 2011), but Paixão et al. (2015) gave that the Gaussian probability distribution function can be applied to create similar settlement profiles as those found in real railway tracks, which can reflect the effect of longitudinal length range and amplitude of settlement simultaneously. Herein, the Gaussian function profiles are used to describe the settlement of ballast subgrade:

$$G(x) = 300 \cdot \alpha \cdot \exp\left[-\frac{(x - x_\mu)^2}{(\beta \times 5.6)^2}\right] / 5.6\sqrt{\pi} \tag{1}$$

where $G(x)$ presents the settlement profiles of ballasted subgrade (unit: mm); x_μ denotes the maximum amplitude of the settlement profile (unit: mm); α and β are constants to obtain many combinations of the amplitude and the longitudinal length for the subgrade settlement, and their value ranges are α = [0, 1] (mm) and β = [0, 1] (m) with the spacing of 0.1 mm/0.1 m, respectively.

Considering the computation efficiency of each numerical step, it is decided to divide amplitude-wavelength domain into ten equally-spaced intervals, thus resulting in 100 settlement profiles by Eq. 1, shown in Fig. 3(a). The range of amplitude is [0, 30 mm], and the extended longitudinal length along the track is [0, 30 m]. The size of settlement zone in the train-track-subsoil model is about 60 m on the right sides, which can be seen in Fig. 2. The application of the profiles to the surface of ballasted subgrade of track model is realized by changing the vertical coordinate of surface nodes (See Fig. 3(b)).

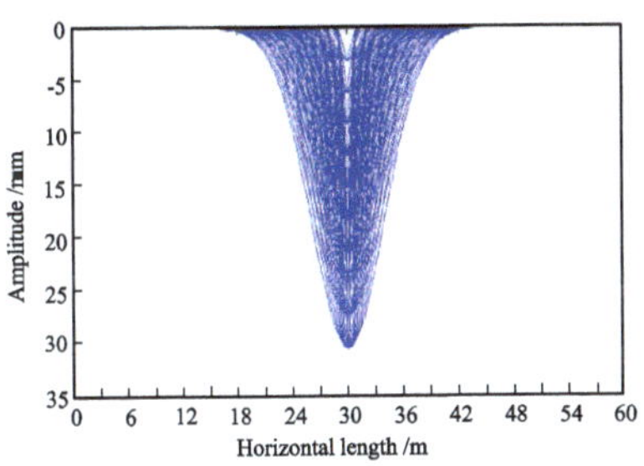

(a) 100 kinds of settlement profiles

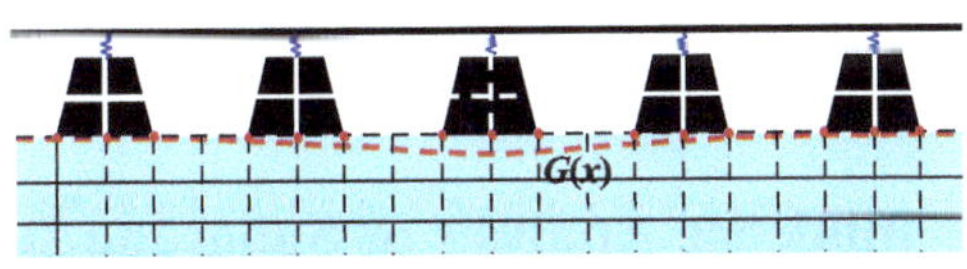

(b) Simulation of settlement in track model

Fig. 3. Settlement profiles and their simulation in track model

3 Validation

The effectiveness of the numerical model can be validated by comparing the results from this paper and those from Paixão et al. (2015), showed in Fig. 4 and Table 3. The results agree very well, which implies the validity of the established numerical model.

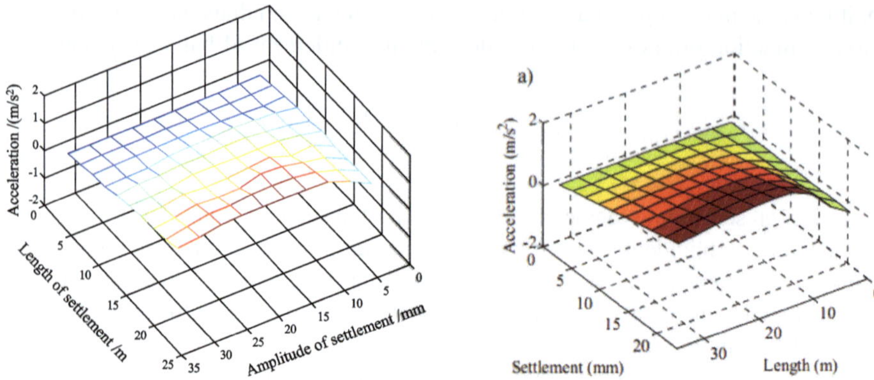

Fig. 4. Comparison of vertical acceleration of car-body (Left: Result by this numerical method, Right: Result from Paixão)

Table 3. Comparison of results by this numerical method and Paixão et al. (2015)

Settlement cases (amplitude/length)	Axle acceleration/ (kg/m^2)		Car-body Acceleration/(kg/m^2)	
	This method	Paixão	This method	Paixão
Case 1 (6.6 mm/6.6 m)	51.19	49.01	0.40	0.42
Case 2 (6.6 mm/9.9 m)	30.34	32.11	0.48	0.51
Case 3 (11 mm/6.6 m)	58.50	57.97	0.65	0.62
Case 4 (11 mm/9.9 m)	71.65	70.23	0.73	0.70

4 Influence Analysis of Subgrade Differential Settlement on Running Trains

4.1 Influence on Vertical Wheel-Rail Interaction Force

When the train runs through the subgrade settlement zone, the vertical wheel-rail interaction forces (w/r forces) are changeable. They may deviate from their normal values, especially in the case of unfavorable combination of settlement amplitude and longitudinal length. Figure 5 gives the variation of the vertical maximum/minimum of w/r forces with 100 kinds of subgrade differential settlements at the train speed of 300 km/h. It can be seen that there are greater variations of w/r forces when the amplitude of settlements are greater and the longitudinal lengths are shorter.

The reason is that the smaller length of settlement means those settlements with short wavelength and the bigger amplitude meaning the bigger vertical deformation of subgrade, and both of them may aggravate the track irregularity. Thus, those subgrade settlements with short wavelength and large amplitude are very disadvantageous to the normal running of high-speed trains.

In order to better observe the dynamic amplification phenomenon of w/r forces, the dynamic amplification factor of wheel-rail force is introduced by $\mu_P = P_D/P_0$, in which

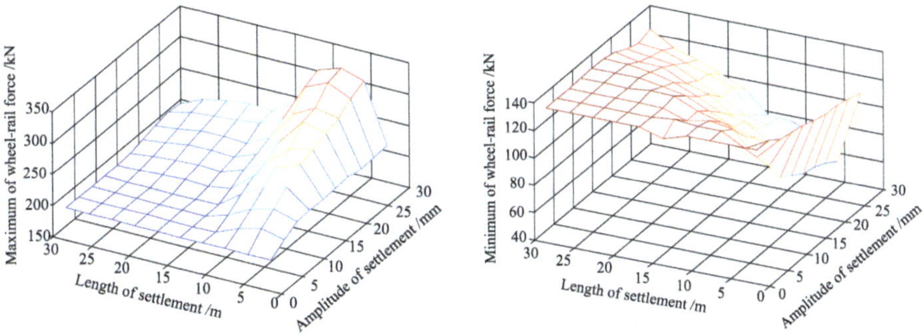

(a) Maximal w/r forces (b) Minimal w/r forces

Fig. 5. Maximal and minimal w/r forces vs settlement

P_D is the peak of wheel-rail interaction force, and P_0 is the static axle-load, and here its average value is about 140 kN for the CRH2 high-speed train.

The variations of μ_P with the subgrade settlement are calculated, as shown in Fig. 6. The contour lines indicate that the dynamic amplification factors are becoming greater and greater with the increasing of the settlement amplitude. When the longitudinal length of settlement zone lies in 5 m–15 m, if the amplitude is higher than 16 mm, the amplification factor μ_P is greater than 1.7; if the amplitude is higher than 24 mm, the amplification factor μ_P is greater than 2.0, which may not only aggravate the degradation of wheelsets and track structures and influence the passenger comfort, but also increase the risk of derailment of running trains.

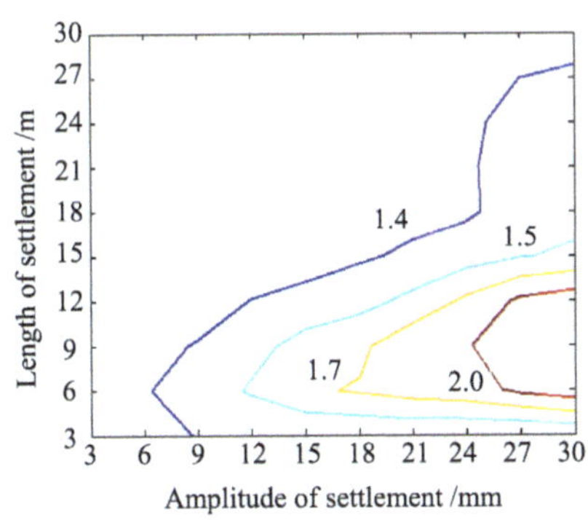

Fig. 6. Contour lines of dynamic amplification factors of w/r forces

4.2 Influence on Passenger Comfort of Trains

The high-speed railways have high requirements for passenger comfort, which is usually evaluated by the riding stability of trains. In Chinese Code for Design of High

Speed Railway (TB10621 2014), the passenger comfort is reflected by the indicator of vertical car-body acceleration. It is regulated that the vertical car-body accelerations are not allowed to be more than 1.3 m/s^2 to guarantee the riding stability and passenger comfort of high-speed trains. Also, when the vertical acceleration of the car-body is less than 1.0 m/s^2, the high-speed railway is regarded to have very good passenger comfort.

The vertical accelerations of the car bodies can be calculated under 100 kinds of subgrade settlements. Figure 7(a) gives the 3-dimensional distribution of vertical acceleration peaks with settlements, at the train speed of 300 km/h. Also, the corresponding contour lines are illustrated in Fig. 7(b). It can be seen from Fig. 7 that: (1) the vertical accelerations of train increase with the increasing of subgrade settlement amplitude, given a certain longitudinal length of settlement; (2) when the longitudinal length of differential settlement is 9 m–27 m, the vertical acceleration may exceed the limit of 1.3 m/s^2 in the case of the amplitude of more than 20 mm; (3) almost for all the longitudinal length of 3 m–30 m of settlement, when the amplitude of differential settlement is more than 9 mm, the passenger comfort may lower.

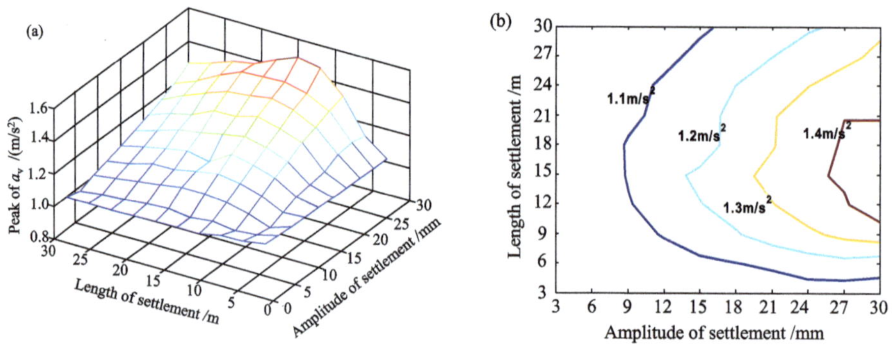

Fig. 7. Vertical acceleration peaks of the car body vs subgrade settlements

The results indicate that the development of subgrade settlement with lengths shorter than 30 m, the amplitudes higher than 20 mm should be strictly limited, and the amplitudes higher than 9 mm should be avoid as soon as possible.

4.3 Influence on Riding Safety of Trains

When the train is running along the track, the wheel loads may be reduced due to train-track vibrations, and many field experiments show that the derailment of running trains is more likely to occur when the wheel load reduction is very large. Therefore, in Chinese Code for Design of High Speed Railway (TB10621 2014), the rate of wheel load reduction is given and defined by $\Delta P/P_0$, where ΔP is the wheel load reduction, and P_0 is the averaged static wheel load; and $\Delta P/P_0 \leq 0.6$ is regulated to guarantee the riding safety of high-speed trains. Figure 8 gives the results of 3-dimensional

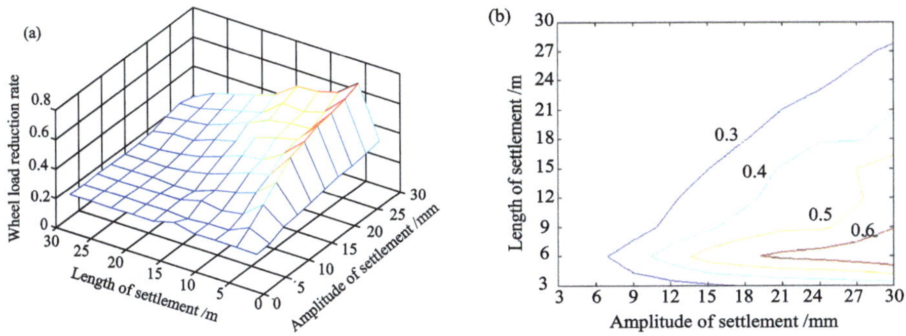

Fig. 8. Peak of wheel load reduction rate vs subgrade settlements

distribution and contour lines of the peak of wheel load reduction rete with the subgrade settlement, respectively, at the train speed of 300 km/h.

One can observe that the development of subgrade settlement with longitudinal lengths of 5 m–9 m, and the amplitudes higher than 19 mm, the wheel load reduction rate is more than the allowance of 0.6, which may increase derailment risk and threaten riding safety of the high-speed train. Therefore, it should be paid more attention to the differential settlements with short lengths and large amplitudes in high-speed railways.

Actually, the riding safety of trains is also evaluated by the derailment coefficient and wheel-rail lateral force. In this section, only the rate of wheel load reduction is used to evaluate the riding safety because the established train-track-subsoil model is two-dimensional.

4.4 Threshold Value of Passenger Comfort and Riding Safety

In order to integrate the analysis of all the numerical results presented above (V = 300 km/h), the image of threshold value is shown in Fig. 9, in which the yellow zone represents poor passenger comfort, the red zone refers to higher derailment risk of running trains, and the purple zone means larger dynamic amplification factors (more than 2.0) and increasing of track degradation.

It can be seen from the threshold image in Fig. 9, there are overlay regions for poor passenger comfort, increasing of track degradation, and higher derailment risk and the unfavorable differential settlements can be found for different performance indicators of trains. Some useful remarks can be drawn: (1) for a certain amplitude of settlement, the shorter is the longitudinal length of settlement, the more damage will occur in the high-speed railway, which accords with the results of Song et al. (2012); (2) those greater amplitudes of settlement are very harmful for the high speed railway, regardless of the longitudinal length of differential settlement; (3) the passenger comfort decreases with the amplitude of the differential settlements, and when the longitudinal length is between 9 m and 27 m, the settlement amplitude should not be more than 20 mm; (4) The riding safety decreases sharply for the settlement length of 5m–9m, when the amplitude of the differential settlements is greater than 19 mm.

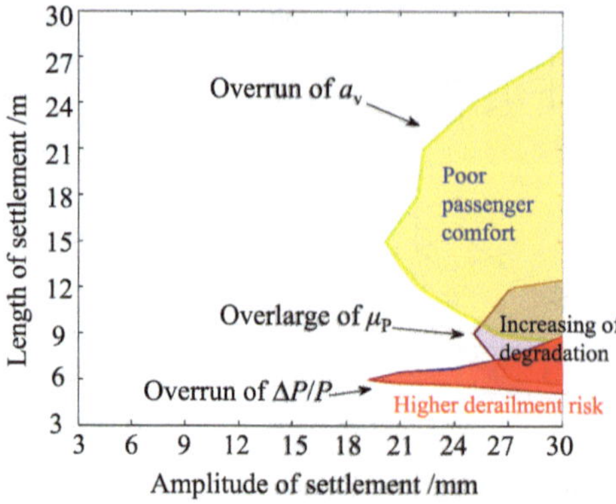

Fig. 9. Threshold of passenger comfort and riding safety vs subgrade settlement

4.5 Influence of Train Speeds

This section aims at the performance indicators under several train speeds of 200 km/h, 250 km/h, 280 km/h, 300 km/h, and 350 km/h, at a definitive subgrade settlement. The peaks of vertical car-body accelerations, the rates of wheel load reduction, and the dynamic amplification factors of wheel-rail forces are calculated and the variation tendency with train speed is showed in Fig. 10. Herein, an unfavorable differential settlement with amplitude of 27 mm and length of 3 m is assigned.

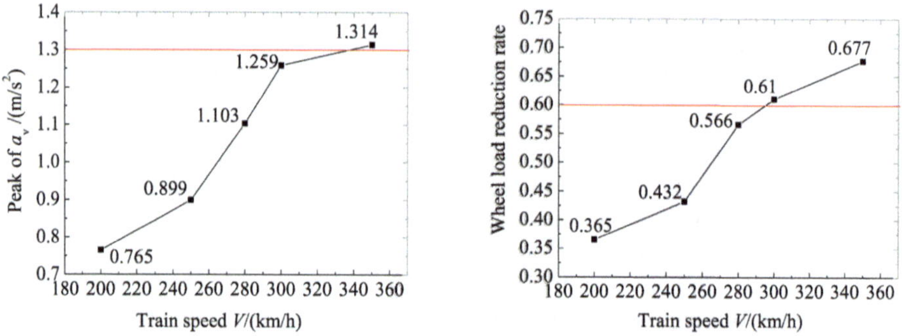

Fig. 10. Car-body accelerations and Wheel load reduction rates vs speed

It can be observed from Fig. 10 that the performance indicators increase obviously with the train speed; the differential settlement of subgrade should be paid more attention to in high-speed railways, because for the same differential settlement of subgrade, the passenger comfort and riding safety can satisfy the allowance at lower train speeds while they cannot at higher speeds.

Therefore, the risk of poor passenger comfort and derailment are much bigger for high-speed railways than for normal trains, and the train speed should be limited in those areas with large differential settlements of subgrade.

5 Conclusions

A dynamic analysis model for train-ballasted track-subgrade system is established to study the effect of different settlement profiles on running properties of high-speed trains in a deeper insight. Due to the complex and computational efforts of 3D model, the 2D model is still useful and applicable in engineering after proper validation. The findings are derived for CRH2 high-speed train in China, but the method is applicable to other cases of train operation and track configurations. In the case study, the dynamic responses of high-speed train and track structure subjected to 100 kinds of differential settlements of subgrade are calculated, and some conclusions can be drawn:

(1) The performance indicators of running trains increase obviously with the train speed, and the differential settlement of subgrade should be paid more attention to in high-speed railways.
(2) The shorter is the longitudinal length of subgrade differential settlement, the bigger the derailment risk in the high-speed railway is.
(3) The greater is the amplitudes of differential settlement, the bigger the probability of poor passenger comfort and higher risk of derailment are.
(4) For the case of CRH2 high-speed train at the speed of 300 km/h, when the longitudinal length is between 9 m and 27 m, the settlement amplitude should not be more than 20 mm, otherwise the passenger comfort cannot be guaranteed.
(5) For the case of CRH2 high-speed train at the speed of 300 km/h, the riding safety decreases sharply for the settlement length of 5 m–9 m, when the amplitude of the differential settlements is greater than 19 mm.

Acknowledgments. Authors wishing to acknowledge assistance or encouragement from colleagues, special work by technical staff or financial support from the Major State Basic Research Development Program of China (“973” Program: 2013CB036203).

References

Abadi, T., et al.: A review and evaluation of ballast settlement models using results from the Southampton Railway Testing Facility (SRTF). Adv. Transp. Geotech. **143**, 999–1006 (2016). The 3rd International Conference on Transportation Geotechnics (ICTG 2016)
Dahlberg, T.: Some railroad settlement models: a critical review. J. Rail Rapid Transit **F4**, 215–289 (2001)
Kang, G.L., et al.: Influence and control strategy for local settlement for high-speed railway infrastructure. Engineering **2**(3), 374–379 (2016)
Knothe, K.L., Grassie, S.L.: Modelling of railway track and vehicle/track interaction at high frequencies. Veh. Syst. Dyn. **22**, 209–262 (1993)

Liu, R.B., Zhang, S.J.: The impact of uneven settlement of embankment to slab track. Subgrade Eng. **1**, 142–143 (2009)

Lysmer, J., Kuhlemeyer, R.L.: Finite dynamic model for infinite media. J. Eng. Mech. Div. **95**, 859–878 (1969)

Mijar, A.R., et al.: Review of formulation for elastostatic frictional contact problems. Struct. Multidiscip. Optim. **20**(3), 167–189 (2000)

Popp, K., et al.: Vehicle-track dynamics in the mid-frequency range. Veh. Syst. Dyn. **31**, 423–464 (1999)

Paixão, A., et al.: The effect of differential settlements on the dynamic response of the train-track system: a numerical study. Eng. Struct. **88**, 216–224 (2015)

Song, H.P., et al.: Correlation between subgrade settlement of high-speed railroad and train operation speed. J. Vib. Shock **31**(10), 134–140 (2012)

Tatsuya, I., et al.: Simple plastic deformation analysis of ballasted track under repeated moving-wheel loads by cumulative damage model. Transp. Geotech. **1**(4), 157–170 (2014)

TB10621: Code for Design of High Speed Railway. Chinese Railway Press, Beijing (2014)

Wang, B.L.: Subgrade Engineering of High Speed Railway. Chinese Railway Press, Beijing (2007)

Xu, Q.Y., et al.: Study on the limited value for the uneven settlement of subgrade under CRTS–I type slab track. China Railw. Sci. **33**(2), 1–6 (2012)

Zou, C.H., et al.: Model test study of influence of differential subgrade settlement on ballasted track settlement. J. Tongji Univ. (Nat. Sci.) **39**(6), 862–869 (2011)

Prediction of Metro Train-Induced Vibrations on a Historic Building: The Case of the Round City and Chengguang Hall in Beijing

Meng Ma[1(✉)], Yanmei Cao[1], Xiaojing Sun[1], and Weining Liu[2]

[1] School of Civil Engineering, Beijing Jiaotong University, Beijing, China
mameng_02231250@163.com
[2] Institute of Rail-System Noise and Vibration Research (IRNVR), Beijing Jiaotong University, Beijing, China

Abstract. With the development of urban mass transit, the train-induced vibrations become a potential problem to historic buildings nearby. The protection and maintenance of historic buildings have aroused a great deal of public attentions. In this paper, the problem of environmental vibrations was analysed for the Round City and Chengguang Hall, a group of historic buildings in Beijing. The historic buildings are located close to a busy road and are near to a planning subway line. The natural frequencies and dynamic amplification factor of the timber structure were estimated by an empirical method. Then, a finite element model was built to predict the metro train-induced vibrations. The results show that, even if the floating slab track was used train-induced vibration can still be a potential problem against long-term protection for the timber structure.

1 Introduction

With the rapid development of urban mass transit system, more and more environmental concerns are focused on the vibrations induced by trains. When considering the long-term vibration effects on historic buildings, the potential architectural damage should be paid more attention. Bata (1971) and Clemente and Rinaldis (1998) reported the damage to some historic buildings caused by road traffic. Ellis (1987) investigated the link between vibration levels arising from road traffic and deterioration of historic buildings. Results showed that architectural damage was often of as much concern as structural damage to heritage buildings. In China, researchers made great efforts to protect historic buildings against vibrations from different types of traffic sources. In 1980s, the plan route of a railway line was changed to protect Longmen Grottoes (Wang and Su 1991). Zhou (2000) investigated the safety problem of Yunyan Pagoda (built between 959 and 961) against vibrations from Beijing-Shanghai high-speed railway lines. Li and Yu (2012) measured the road traffic-induced vibrations in the Hall of the Holy Mother (built between 1023 and 1032) and suggested a minimum safety distance from the road to protect the statues inside.

To learn about the dynamic behaviour of a building, the modal analysis is useful which can be performed by measurement and finite element (FE) analysis. Bazaco et al.

J. Pombo and G. Jing (eds.), *Recent Developments in Railway Track and Transportation Engineering*, Sustainable Civil Infrastructures, DOI 10.1007/978-3-319-61627-8_10

(1995) investigated the behaviour of XVth century Palacio de Sta. Cruz under road traffic loads. Bongiovanni et al. (2011) resumed some experiences in the field of experimental dynamic characterization of historical buildings.

To evaluate a historic building under the vibration environment, the in-situ measurement is the mostly used approach. Meng et al. (2008, 2009) performed experimental investigations of vibration induced by road traffic and evaluated the impact on an ancient city wall and the Bell Tower in Xi'an. Li et al. (2010) measured the metro-induced vibration effect on two sites of surrounding historic buildings. Min et al. (2013) performed an ambient vibration test on a heritage timber building in Seoul. Hinzen (2014) measured the subway train-induced vibrations in Cologne Cathedral.

To evaluate a historic building near a new built railway or subway lines, vibration prediction is an essential work. Most predictions were based on the numerical analysis. Liu et al. (2007) and Jia et al. (2008) predicted the vibration impact on the city wall and a historic railway station above a new-built underground railway line in Beijing. Breccolotti et al. (2011) predicted the vibration levels produced by railway traffic based on experimental measurements and FE modeling. Ma et al. (2016) analysed the vibration impact on a historic Bell Tower in Xi'an above two spatially overlapping metro lines.

Generally speaking, it is not difficult to build FE models for masonry structures. Nevertheless, it becomes a problem in dealing with the traditional Chinese timber structures. Firstly, all the wooden elements are constructed by the tenon-and-mortise work. In addition, most beams and columns are connected by *dougong* which is a unique structural element of interlocking wooden brackets, one of the most important elements in traditional Chinese architecture. Moreover, all columns just stand on the masonry bases, and the horizontal restraint for the columns is only provided by the friction between columns and bases. It can be found that the input parameters for contact element are difficult to determine. Accordingly, under all the difficulties, FE models may be not a good approach to deal with the prediction work. To learn the dynamic behavior of such unique timber structures, the empirical estimation or the in-situ measurement are another alternative choices.

In this paper, the prediction of metro train-induced vibrations on Chengguang Hall, located on the Round City, was studied. Firstly, the dynamic characteristic of the timber structure was estimated by an empirical method. Then, two 3D finite element (FE) models for calibration and prediction were built. Finally the vibration responses were analysed.

2 Problem Outline

2.1 Introduction of the Historic Building

Chengguang Hall (the Hall of Receive Light) was originally built in 1264 and reconstructed in 1743. As a typical Chinese timber structure, it is a spacious building with a double-eaved roof made of yellow glazed titles bordered in green (Fig. 1). Chengguang Hall is the main structure on the Round City, a mini castle with the area of only 4,553 m^2 and a round brick wall with about 4 m height on a island in the south of

Fig. 1. A picture and the cross-section of Chengguang Hall

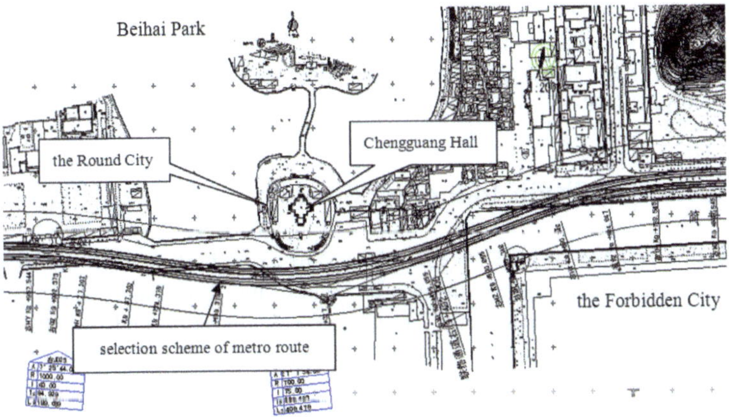

Fig. 2. Layout of planned metro route near the Round City

Beihai Park (Fig. 2). Together with Beihai Park, Chengguang Hall and the Round City were listed as the first batch of national key cultural relics protection units of China in 1961.

2.2 Estimation of Dynamic Characteristic

In order to learn about the dynamic characteristic of a historic building, three methods can be employed: in-situ measurement, numerical simulation and empirical estimation. Here an empirical method was used according to Chinese national code GB/T50452 (2008). The first, second and third natural frequencies f_i can be estimated by:

$$f_i = \frac{1}{2\pi H} \lambda_j \psi \tag{1}$$

where H is the total height of a core column; ψ is a stiffness factor, and for a timber frame hall built on a massive platform foundation, a statistical value suggests $\psi = 43$ m/s; λ_j is a calculation factor for the j-th natural frequency, which depends on the structural dimension and can be determined by an empirical table from GB/T50452 (2008). According to the surveying and mapping of Chengguang Hall by Ma et al. (1987), the calculation factors can be determined as $\lambda_1 = 1.79$, $\lambda_2 = 4.76$ and $\lambda_3 = 7.76$, and the natural frequencies can be calculated as f_1 = 1.16 Hz, f_2 = 3.09 Hz, f_3 = 5.03 Hz.

The horizontal maximum peak particle velocity (PPV) of a timber structure can be estimated by

$$V_{\max} = V_r \sqrt{\sum_{j=1}^{3} \left(\gamma_j \beta_j\right)^2} = V_r D \quad (2)$$

where V_r is the horizontal PPV at the foundation of the building, γ_j and β_j are the participating factor and dynamic amplifying factor of the *j*-th vibration mode, respectively. *D* is the total dynamic amplification factor, which can be determined as 2.565 by empirical table from GB/T50452 (2008).

3 Prediction of Train-Induced Vibration

As a selection scheme of a metro route is planed to be built near to the Round City (Fig. 2). To predict the vibration responses of both the Round City and Chengguang Hall, two 3D FE models were built - one used for calibration and the other used for prediction. The prediction model includes the twin tunnels, soil layers, the Round City and the foundation of Chengguang Hall. Nevertheless, the timber structure of Chengguang Hall was not taken into account, due to the difficulties to model the timber structures which mentioned in introduction. Instead, the estimated dynamic amplification of the structure was used here.

3.1 3D FE Models

Two 3D FE models were built (Fig. 3). One was used for calibration, whose model dimension and input soil parameters were from Beijing metro line M4. The other was used for prediction, whose input soil parameters were from the core drilling data near the Round City. The input wave velocities for both models were plotted in Fig. 4. In the Beijing M4 where the calibration measurement was performed, the train speed is approximately 40 km/h, and the floating slab track (FST) with steel springs is installed. In the planned metro line near the Round City, the train speed and the parameters of FST keep the same. A train-FST track model was built, from which the supporting forces of the steel springs were obtained, and then applied on the concrete invert in tunnels.

3.1.1 Simulation of Moving Train Loads

To calculate the moving train loads, a MATLAB based program STFSTI (Simulation of Train/Floating Slab Track Interaction) was used. This program was developed by

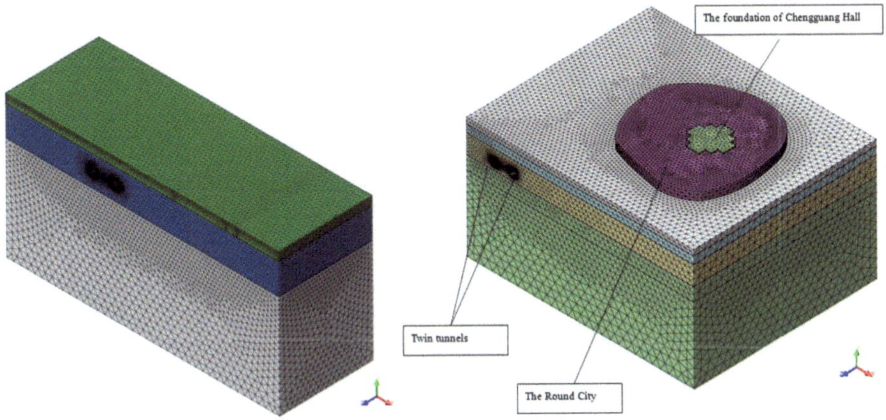

Fig. 3. Calibration model (left) and prediction model (right)

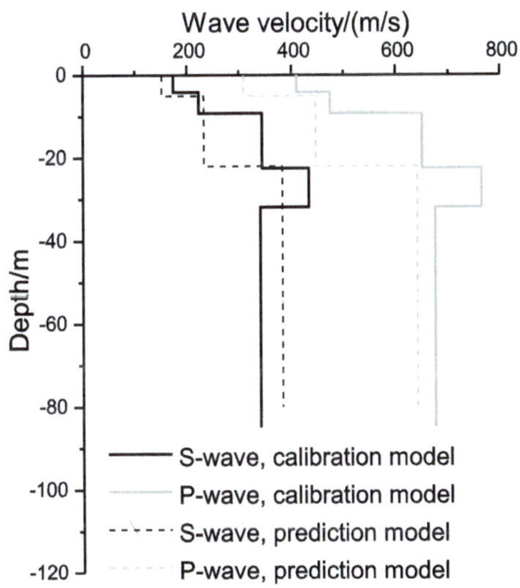

Fig. 4. The input wave velocities for FE models

Beijing Jiaotong University (BJTU) (Ma 2014; Liu et al. 2016). The modeling in STFSTI is restricted in two dimensions. The vehicles are modeled by a mass-spring system that consists of one car body, two bogies and four wheels, all of which are rigid bodies and connected to each other by the secondary and primary suspensions. The floating slab track model is based on periodic-infinite structure theory. The rail is modeled by an infinite Euler beam, and the floating slab is modeled by Euler beam with finite length. Within each length of a floating slab, there are N_r fasteners and N_s steel springs, with a distance of d_r and d_s (Fig. 5). The calculation of this program is performed in frequency domain and the output is analytical solution. The contact wheel/rail forces are modeled by a Hertzian contact spring.

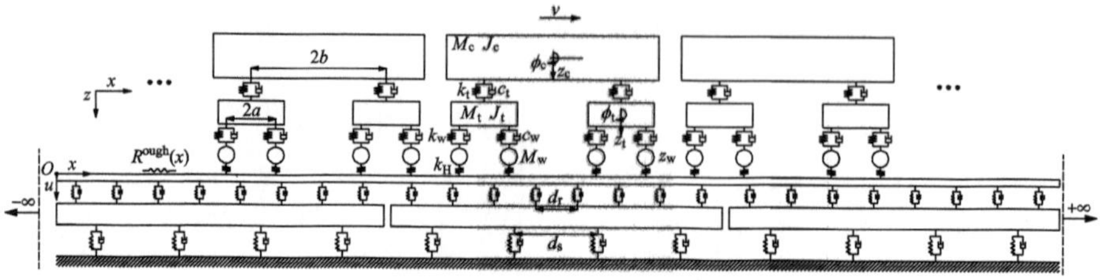

Fig. 5. Schematic of dynamic train-track model

The dynamic forces under the p-th steel spring of j-th floating slab can be calculated by:

$$\hat{F}_{j,p}(\omega, \omega_l) = \frac{(k_s + ic_s\omega)}{2}\hat{u}_s(x_{j,p}, \omega, \omega_l) \tag{3}$$

where $\hat{u}_s(x_{j,p}, \omega, \omega_l)$ is the displacement response at the p-th steel spring of j-th floating slab; $x_{j,p}$ is the coordinate of the $\hat{u}_s(x_{j,p}, \omega, \omega_l)$; k_s and k_p are the stiffness and damping of the spring, respectively.

According to the periodic-infinite structure theory, the displacement can be expressed as:

$$\hat{u}_s(x_{j+1,p}, \omega, \omega_l) = e^{i(\omega_l - \omega)L/v}\hat{u}_s(x_{j,p}, \omega, \omega_l) \tag{4}$$

Then, under the wheel/rail forces $\bar{Q}_k^{wr}(\omega_l) = e^{it\omega_l}$ (k = 1, 2, …, m_w), the forces are also abided by the periodic condition:

$$\hat{F}_{j+1,p}(\omega, \omega_l) = e^{i(\omega_l - \omega)L/v}\hat{F}_{j,p}(\omega, \omega_l) \tag{5}$$

Finally, by the principle of superposition, the total force can be expressed as:

$$\hat{F}_{j,p}(\omega) = \sum_{l=0}^{N_R}\hat{F}_{j,p}(\omega, \omega_l) \tag{6}$$

3.1.2 In-Situ Measurement and Model Verification

Figure 6 shows the comparison of measurement and calculation results, in time history and in one-third octave bands. The measurement results are chosen from BJTU's database of railway vibration, which is the vertical ground velocity 30 m away from the central line of the shield tunnel of M4. The train speed was approaching 40 km/h and the track is steel spring FST.

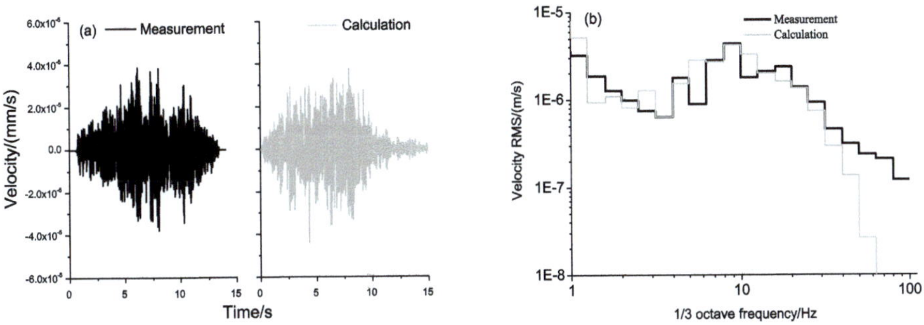

Fig. 6. Comparison of measured and calculated results: (a) time history; (b) 1/3 octave band

3.2 Prediction of Vibration Responses

Two conditions are considered herein: one is a train running in the tunnel near the Round City, and the other is two trains running in both tunnels at the same time. Figure 7 shows the vertical velocities of the foundation of Chengguang Hall and the Round City when the two trains operate in the both tunnel at the same time. It shows the maximum PPV is 0.146 mm/s. Besides, on the top of the city wall, vibrations do not attenuate monotonously with the distance to the source. The existence of the foundation affects the vibration field around it.

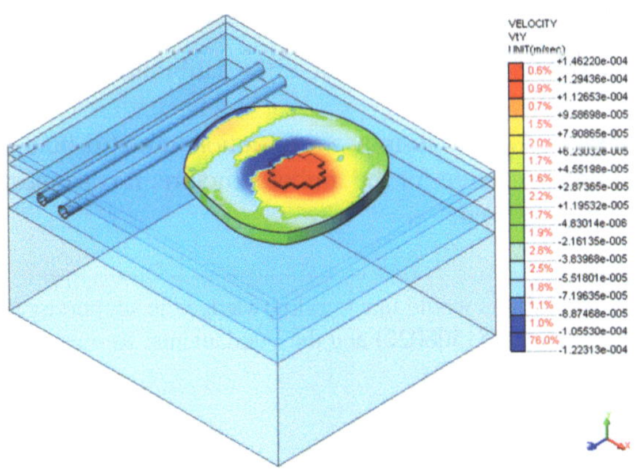

Fig. 7. Vertical velocity of the foundation of the hall and the Round City

Figure 8 shows the predicted PPVs. According to GB 50452, the guidance limit for the Round City and Chengguang Hall are 0.20 and 0.15 mm/s, respectively, considering the building protection against architectural damages. When predicting the vibration of timber structure, the total dynamic amplification factor *D* is introduced. It can be found that, the vibration response can satisfy the protection limit of the Round City. Nevertheless, the horizontal vibration of the timber structure exceeds the limit

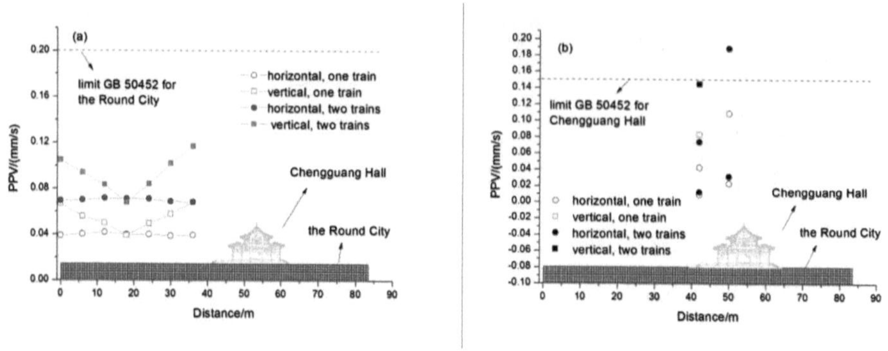

Fig. 8. Vibration responses of (a) the Round City and (b) Chengguang Hall

under condition of two-train operation. In reality, the predicted responses are only the vibration induced by metro trains. If the contribution of road traffic is considered, the total responses will be larger.

4 Conclusions

In this study, a case study of train-induced vibration impact on historic buildings was introduced. By the empirical estimation of the dynamic behavior of the timber structure and numerical prediction, the following conclusions can be drawn:

(a) By comparison of the ground vibration of the calibration model, the FE model can be accepted;
(b) Even if the FST is used, train-induced vibration can still be a potential problem against long-term protection for the timber structure; it is suggested that the metro route should be optimized.

Acknowledgments. The authors would like to acknowledge the support by the National Science Foundation of China (Grant no. 51508022) and the Fundamental Research Funds for the Central Universities of China (2016JBM046).

References

Bata, M.: Effects on buildings of vibrations caused by traffic. Build. Sci. **6**, 221–246 (1971)
Bazaco, M., Montoya, F., Alvarez, V., et al.: Traffic induced vibrations in historic buildings. Case of study: Palacio de Sta. Cruz of Valladolid. WIT Built Environ. **15**, 109–118 (1995)
Bongiovanni, G., Clemente, P., Rinaldis, D., et al.: Traffic-induced vibrations in historical buildings. In: Proceedings of the 8th International Conference on Structural Dynamics (EURODYN 2011), Leuven, pp. 812–819 (2011)

Breccolotti, M., Materazzi, A.L., Salciarini, D., et al.: Vibrations induced by the new underground railway line in Palermo, Italy - experimental measurements and FE modeling. In: Proceedings of the 8th International Conference on Structural Dynamics (EURODYN 2011), Leuven, pp. 719–726 (2011)

Clemente, P., Rinaldis, D.: Protection of a monumental building against traffic-induced vibrations. Soil Dyn. Earthq. Eng. **17**(5), 289–296 (1998)

Chinese National Code, GB/T50452: Technical specifications for protection of historic buildings against man-made vibration. China Building Industry Press, Beijing (2008)

Ellis, P.: Effects of traffic vibration on historic buildings. Sci. Total Environ. **59**(4), 37–45 (1987)

Hinzen, K.G.: Subway-induced vibrations in Cologne Cathedral. Seismol. Res. Lett. **85**(3), 631–638 (2014)

Jia, Y.X., Liu, W.N., Liu, W.F., et al.: Study of vibration effects on historic buildings due to moving trains in Beijing. In: 9th International Symposium on Environmental Geotechnology and Global Sustainable Development, Hongkong, pp. 492–499 (2008)

Li, K.F., Liu, W.N., Liu, W.F., et al.: Tests and analysis of metro-induced vibration effects on surrounding historic buildings. In: 1st International Conference on Railway Engineering: High-Speed Railway, Heavy Haul Railway and Urban Rail Transit, Beijing, pp. 921–926 (2010)

Li, Q.L., Yu, Z.L.: Influence of automobile-induced vibration on the colored Song Dynasty clay statues in the Sacred Lady Hall. Sci. Conserv. Archaeol. **24**(1), 49–55 (2012). (in Chinese)

Liu, W.F., Liu, W.N., Jia, Y.X., et al.: Study on effect on ming dynasty's city wall due to train induced vibrations in Beijing. In: 3rd International Symposium on Environmental Vibration, Taipei, pp. 229–234 (2007)

Liu, W.N., Ma, L.X., Jiang, B.L., et al.: Generalized wavenumber method for dynamic response analysis of floating slab track. China Railw. Sci. **37**(1), 31–38 (2016)

Ma, B.J., et al.: Surveying and mapping for Chengguang Hall on the Round City of Beihai Park. Tradit. Chin. Archit. Gard. **3**, 48–55 (1987). (in Chinese)

Ma, M., Liu, W.N., Qian, C.Y., et al.: Study of the train-induced vibration impact on a historic Bell Tower above two spatially overlapping metro lines. Soil Dyn. Earthq. Eng. **81**, 58–74 (2016)

Ma, L.X.: Study on the model of coupled vehicle & track and the analysis model for tunnel-ground vibration response based on the periodic-infinite structure theory. Ph.D. thesis, Beijing Jiaotong University, Beijing (2014)

Meng, Z.B., Chang, Y.Z., Song, L., et al.: The effects of micro-vibration excited by traffic vehicles on Xi'an Bell Tower. In: The 2nd International Conference of Transportation Engineering (ICTE 2009), Chengdu (2009)

Meng, Z.B., Hu, W.B., Wu, M.Z., et al.: An experimental investigation of micro-vibration of Xi'an ancient city wall excited by traffic vehicles. In: 10th International Symposium on Structural Engineering for Young Experts, Changsha, pp. 1129–1134 (2008)

Wang, Z.L., Su, G.: Environmental impact on Longmen Grottoes in Luoyang by Jiaozhi Railway. Railw. Stand. Des. **12**, 35–39 (1991). (in Chinese)

Zhou, J.H.: Study of propagate law of vibration caused by train on the high speed railway. In: Mechanics 2000, Beijing, pp. 659–661 (2000). (in Chinese)

Performance of Polyurethane Polymer in the Transition Zones of Ballasted and Ballastless Track

Qi Wei[1(✉)] and Li Cheng-hui[2]

[1] Key Laboratory of High-Speed Railway Engineering, Ministry of Education, Southwest Jiaotong University, Chengdu 610031, China
qw9199@126.com
[2] School of Civil Engineering, Southwest Jiaotong University, Chengdu, China

Abstract. Track disease significantly increased in the transition zones between ballasted and ballastless track. It is triggered by abrupt variation in the track's vertical stiffness due to different support conditions. To achieve the transition of track stiffness, the ballast bed with polyurethane polymer scheme was proposed. In this scheme, the changed of ballast bed stiffness change quantity was decided by the bonding area of ballast bed with polyurethane polymer. In the test section, three bond forms of ballast bed were adopted—full bonding section, partial bonding section and local bonding section. The field test results were as follows: the vertical stiffness of ballast bed with polyurethane polymer was increased more than 5 times, 3 times, 2 times. The vertical and horizontal resistance of ballast bed was increased more than 8 times, 4 times, 3 times. With the increase of track stiffness, the vertical force of wheel-rail and the vertical vibration of rail increase, the vertical displacement of rail reduce, the vertical displacement and vertical vibration of sleeper reduce. Merely citing the vertical displacement of rail as typical, it was reduced by 25%, 16%, 3%. After the static and dynamic test, we can conclude that performance of polyurethane polymer in the transition zones can achieve the transition of ballast bed from low stiffness to high stiffness.

1 Introduction

Track diseases in the transition of ballasted and ballastless track are significantly increased, such as sleeper damage, ballast crusher, slab cracking, etc. It is caused by abrupt variation in the track's vertical stiffness due to different support conditions [1–3]. Heavy haul railway in China mainly used in ballasted track, but its standard provides that ballastless track need to be used in the long tunnel and set transition section of track structure between ballasted track and ballastless track. In China, a large number of railway lines have been build and planned, and the transition zones appeared inevitably. In order to reduce the track diseases, it need to set a proper track transition section.

There are many ways to improve transition zones of ballasted and ballastless track, such as adding auxiliary rail, widen and extended sleeper, decreases sleeper spacing, increasing the thickness of ballast, etc. [4]. The change of track structure affected the

J. Pombo and G. Jing (eds.), *Recent Developments in Railway Track and Transportation Engineering*, Sustainable Civil Infrastructures, DOI 10.1007/978-3-319-61627-8_11

maintenance operations, because of the widely application of standard mechanical equipment in Chinese railway line, then these treating measures have been rarely used. At present, ballast stiffness transition and fastener stiffness transition have been applied gradually, due to its simple manipulation and easy change of the track stiffness.

Asphalt, epoxy resin and polyurethane polymer can change ballast stiffness. Asphalt has low heat stability and requires high foundation drainage, so it is gradually being eliminated. Epoxy resin has poor impact resistance and easy to crack, so it is not application. Polyurethane polymer has good elasticity, toughness and plastic, and it can change ballast stiffness by control adhesive section of ballast bed, so it is gradually being applied.

The application of polyurethane polymer has been used in United Kingdom, Germany, China, etc. Since 1990s, Peter. K. Woodward has been carried out XiTRACK Technique, and conducted indoor and field testing. XiTRACK is a patented technique that has been designed to enhance the engineering properties of the railway track. It has been successfully applied to reinforce the track bed's stabilization on the West Coast Main Line near Bletchley, and fixed geometry of ballasted in tunnels, high-speed turnouts, bridge transitions, crossings and turnouts. After applied XiTRACK, line disease problems have been effectively controlled for more than 10 years [5–9]. In China, MOE key laboratory of high-speed railway engineering has done much performance of polyurethane polymer indoor testing, and have introduced a standard [10]. Part of the materials property refer to Tables 1, 2 and 3.

Table 1. Performance parameters

Test item	Performance index
Viscosity (MPa·s)	500–1500
Time of surface drying (h)	≤ 1
The polymerization shrinkage mass relate (mm)	≤ 3

Table 2. Mechanical property

Test item		Performance index
Apparent density (g/cm^3)		1.15 ± 0.05
Shore hardness of A type (°)		≥ 90
Bond strength (MPa)		≥ 2 (或底材开裂)
Tensile strength (MPa)		≥ 13
Elongation at break (%)		≥ 250
Low-temperature brittleness (°C)		≤ −50
Compressive strength (MPa)		≥ 40
hot air aging (168 h, 90 °C)	Retention percentage of tensile strength	≥ 70%
	Retention percentage of elongation at break	≥ 70%

(continued)

Table 2. (*continued*)

Test item		Performance index
Damp heat aging (80, 95% RH, 168 h)	Retention percentage of tensile strength	≥70%
	Retention percentage of elongation at break	≥70%
Ultraviolet aging (168 h)	Retention percentage of tensile strength	≥70%
	Retention percentage of elongation at break	≥70%

Table 3. The indoor test results of ballasted track supporting stiffness with different cured thickness

Cured thickness (total thickness is 350 mm)	Primary ballast	Super ballast
200	120	130
250	140	160
350	200	240

Track disease in the transition zones between ballasted and ballastless track is triggered by abrupt variation in the track's vertical stiffness due to different support conditions. XiTRACK Technique can improve the strength and stability of the ballast bed, and it effectively reduced the ballast accumulated deformation, but there is less research in this field. Therefore, it is used for bonding ballast in the transition sections of ballasted and ballastless track to reduce the volume of maintenance, and improve the transport ability, traveling comfort and safety. Up to now, the test section of the application of Polyurethane polymer in the transition of ballasted and ballastless track was established in Datong-Xian line, Haerbin-Dalian line, Chengdu-Qingchengshan line, Shanxi Mid-south-part heavy rail line, etc. (in China), and this test section is located in Shanxi Mid-south-part heavy haul rail line.

Polyurethane polymer in the Transition Zones can be achieved the stiffness transition of ballastless track to ballasted track in theory, so we established the test section to verify it by field test. Field test can further optimize the transition plan, and provide technical support for its large scale promotion. (The flow chart of this paper is shown in Fig. 1).

2 Engineering Overview

This test section is located at the entrance of the Hongling tunnel. The length of the ballast bed with polyurethane polymer is 20 m, and they all in the tunnel, see Fig. 2. In order to achieve a gradual transition of the ballasted bed stiffness, the bonding section of the ballasted bed with polyurethane polymer are respectively the full bonding

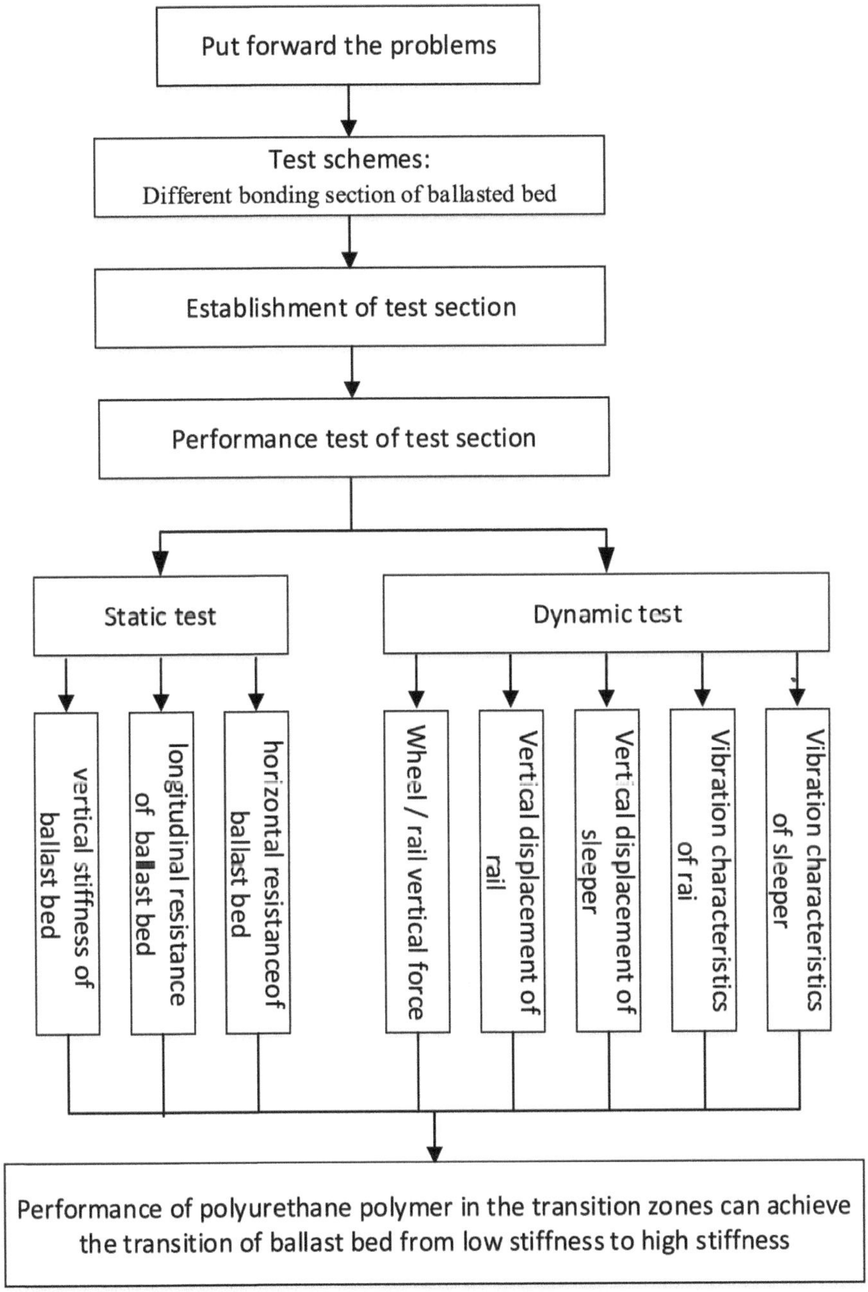

Fig. 1. The flowchart

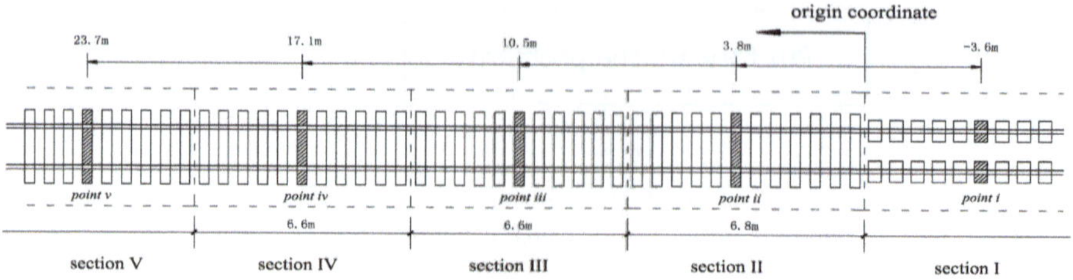

Fig. 2. General layout of the site

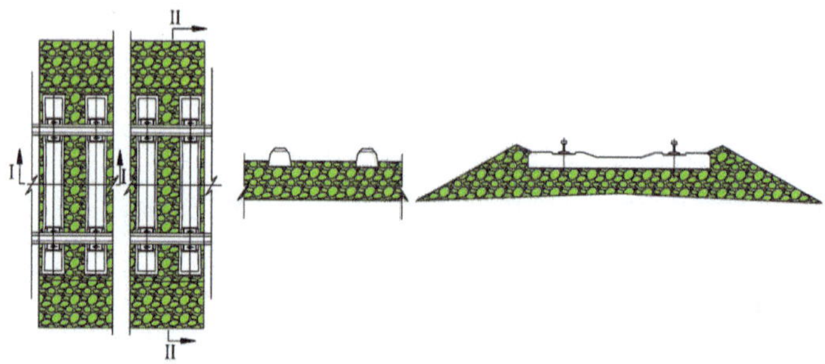

(a) plane graph (b) skiagraphy section of I-I (c) drawing cross section of II-II

Fig. 3. Full bonding section

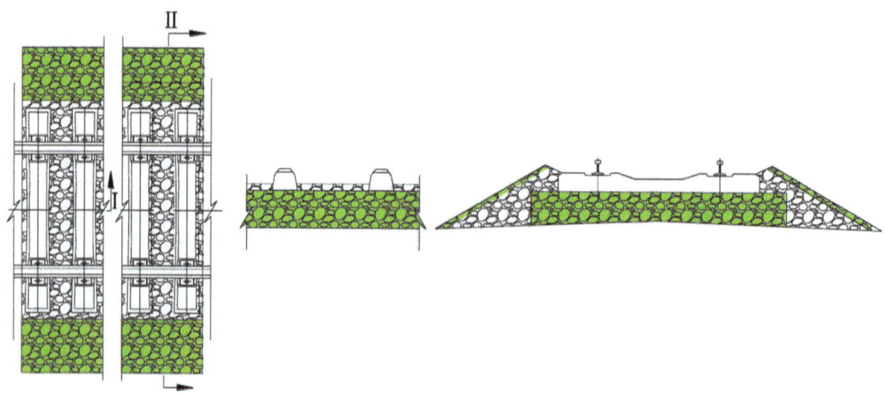

(a) plane graph (b) skiagraphy section of I-I (c) drawing cross section of II-II

Fig. 4. Partial bonding section

section in Fig. 3 (II section in Fig. 2), partial bonding section in Fig. 4 (III section in Fig. 2) and local bonding section in Fig. 5 (IV section in Fig. 2). Section I is low vibration track with ballastless bed, and Sect. II is the ballasted track.

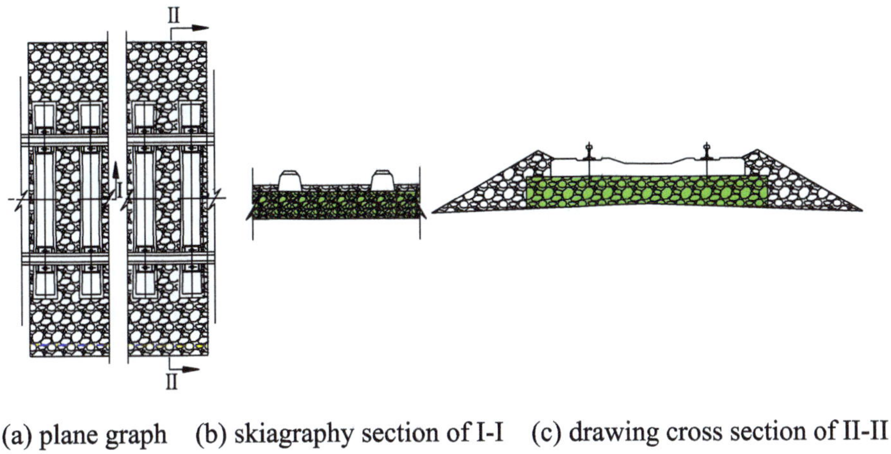

Fig. 5. Local bonding section

3 Test Equipment

The performance of ballast bed was significantly changed after the ballast have been bonded. This test is divided into static test and dynamic test. The static test include the vertical stiffness and longitudinal resistance and horizontal resistance of ballast bed. The dynamic test include the wheel/rail vertical force, the vertical displacement of rail and sleeper, vertical vibration of rail and sleeper.

Static testing instruments are our own design and production. The instruments of ballast bed's vertical stiffness is shown in Fig. 6. In the static test, load was applied by a hydraulic jack, Displacement was tested by the dial gauge.

The dynamic testing instruments include data acquisition equipment, strain gauge, accelerometer, displacement gage, etc. The data acquisition equipment is produced by the company of Integrated Measurement & Control Co, and a CX_5032 instrument was used for the acquisition of wheel/rail force and vertical displacement, and two CS_3008 instruments were used for the acquisition of vertical vibration. Domestic strain gauges which resistance value is 120 Ω are used in wheel/rail force and displacement test (in Fig. 7), and it was connected with full bridge circuit by the Reference of Test method for horizontal force and vertical force of wheel rail (in China).

The test train's height is 5000 t, and its axle load is 30 t. The speed of test train were 60 km/h, 70 km/h, 80 km/h, 90 km/h, and the train ran back and forth many times.

The test section takes the junction of ballasted and ballastless track as the origin of coordinate, and the direction to the ballasted is designated as the positive direction. Data acquisition points were point *i* at −3.6 m, *ii* at 3.8 m, *iii* at 10.5 m, *iv* at 17.1 m, *v* at 23.7 m, as shown in Fig. 2. Because of the existence of discrete in static test, a plurality of sleepers were tested in test section of ballasted bed which have been bonded.

Fig. 6. Test of ballast bed's vertical stiffness

Fig. 7. Dynamic test

4 Analysis of Static Test

The vertical stiffness of ballast bed is shown in Fig. 8, in comparison to the vertical stiffness of ballast bed which apply polyurethane polymer before and after, it increased more than 5 times, 3 times, 2 times at section. II, section III, section IV. It can be concluded that the bonded ballast which under the sleeper played a major role in enhancing the vertical stiffness of ballast bed, and secondly was the bonded ballast which between the sleeper and the shoulder of ballasted bed.

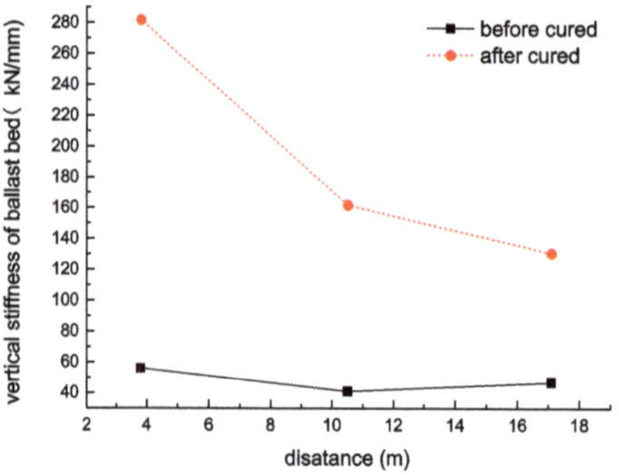

Fig. 8. Result of ballast bed's vertical stiffness

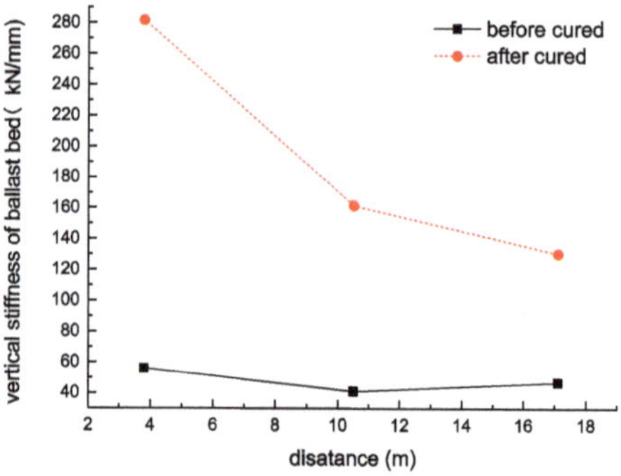

Fig. 9. Result of ballast bed's longitudinal resistance

The longitudinal and horizontal resistance of ballast bed is shown in Figs. 9 and 10, its value of ballast bed which apply polyurethane polymer increased more than 8 times, 4 times, 3 times. This is because the bonded ballast which under the sleeper had played the most important par for the increase of ballast bed's resistance, next was the bonded ballast which between the sleeper, and then was the bonded ballast which at the shoulder of ballasted bed.

Test results show that the performance of ballast bed which apply polyurethane polymer has caused huge changes, but the amount change are not the same at different bonding section of the ballasted bed. Performance of polyurethane polymer in the transition zones achieved the transition of ballast bed from low stiffness to high stiffness.

5 Analysis of Dynamic Test

5.1 Wheel/Rail Vertical Force

The test results as shown in Fig. 11, the wheel/rail force is gradually decreased from section II to section IV, and the value of section II is the largest, and section I is the smallest. Taking the force at speed 80 km/h as an example, the force of section II to section IV increased by 8.05%, 3.94%, 0.97% than section V. The wheel/rail force is influenced by the track stiffness, so we know that the track stiffness gradually decreased from section II to section V. The ballastless track is located in section I, its track stiffness is similar to the ballasted track due to the coordination of the stiffness of fastener and ballast bed. While, at section II, section III, section IV, their ballast bed stiffness increases, and the faster stiffness has not changed because they used the faster for ballasted track, so their wheel/rail forces are larger than section I.

From Fig. 11, we can also know that, the wheel/rail vertical force gradually increased with the increase of vehicle speed. Taking the force at section II as an example, the force increased by 4.6%, 12.9%, 22.4% from speed 60 km/h to 90 km/h.

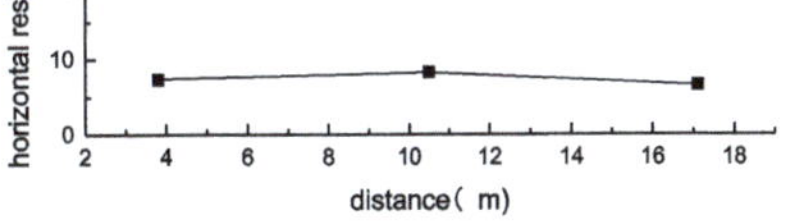

Fig. 10. Test result of ballast bed's horizontal resistance

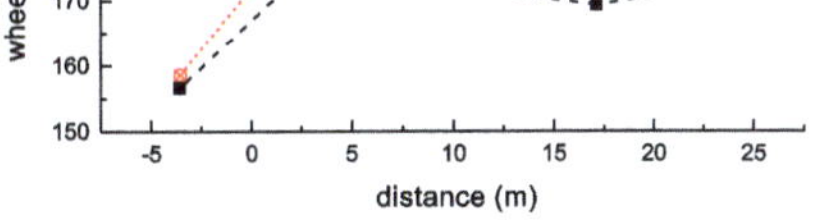

Fig. 11. Wheel/rail vertical force

5.2 The Vertical Displacement of Rail and Sleeper

The vertical displacement of rail is shown in Fig. 12, its value reduce at the ballast bed which apply polyurethane polymer. The vertical displacement of rail gradually increased from section II to section IV, taking the vertical displacement at speed 80 km/h as an example, their vertical displacement are 0.75 times, 0.84 times, 0.97

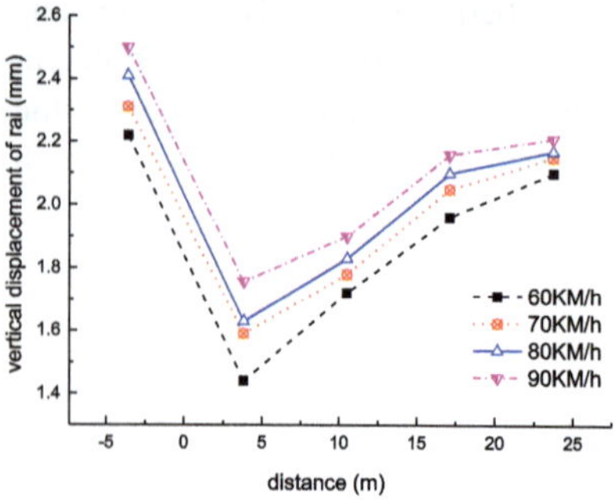

Fig. 12. Vertical displacement of rail

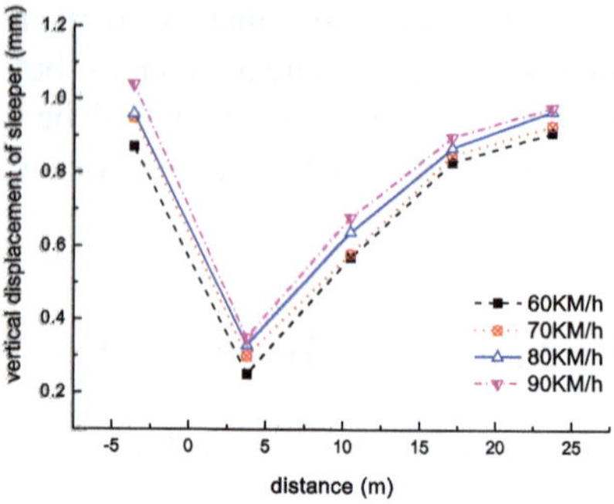

Fig. 13. Vertical displacement of sleeper

times of section V. The vertical displacement of rail gradually increased with the increase of vehicle speed. Taking the force at section III as an example, the vertical displacement increased 10% from speed 60 km/h to 90 km/h.

The vertical displacement of sleeper is shown in Fig. 13, its value reduce at the ballast bed which apply polyurethane polymer, especially at section II. The vertical displacement of sleeper at section II only 27.5% of the V section, this is because the full bonding section of ballasted bed had bonded the sleeper and ballast bed as a whole, and the ballast stiffness is greatly improved. Section I is Low Vibration Track, the rubber pad is located under the block-supported, so, its vertical displacement is relatively large. The vertical displacement of sleeper gradually increased with the increase of vehicle speed.

5.3 Vertical Vibration of Rail and Sleeper

The vertical vibration of rail is shown in Fig. 14, it is gradually decreased from section II to section IV. Because after the ballast bed applied polyurethane polymer, its integrity has been improved, and then the vertical vibration of ballast bed reduced, so the vertical vibration of rail increased. Because section V is located outside of the tunnel, its vertical vibration is different from other sections. The vertical vibration of rail gradually increased with the increase of vehicle speed. Taking the force at section III as an example, vertical vibration of rail increased 95% from speed 60 km/h to 90 km/h.

The vertical vibration of sleeper is shown in Fig. 15, it is gradually increased from section II to section V. The vertical vibration at section I is larger than section II, because after the ballast bed applied polyurethane polymer, its integrity has been improved, and then its vertical vibration reduced. While, at section I, the rubber pad is located under the block-supported, its vertical vibration relatively larger. The vertical vibration of sleeper gradually increased with the increase of vehicle speed. Taking the force at section III as an example, vertical vibration of rail increased 94% from speed 60 km/h to 90 km/h.

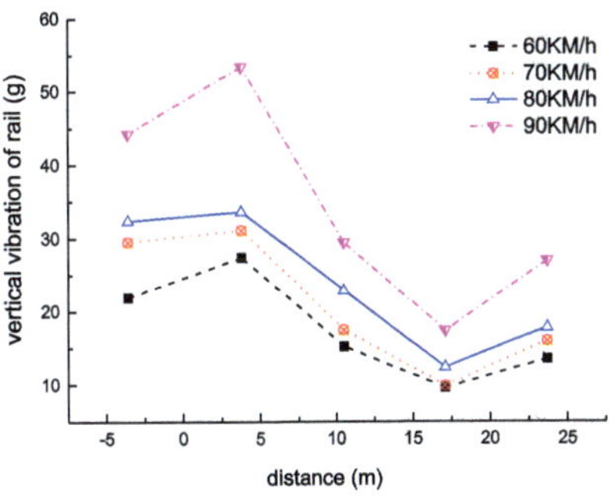

Fig. 14. Vibration characteristics of rail

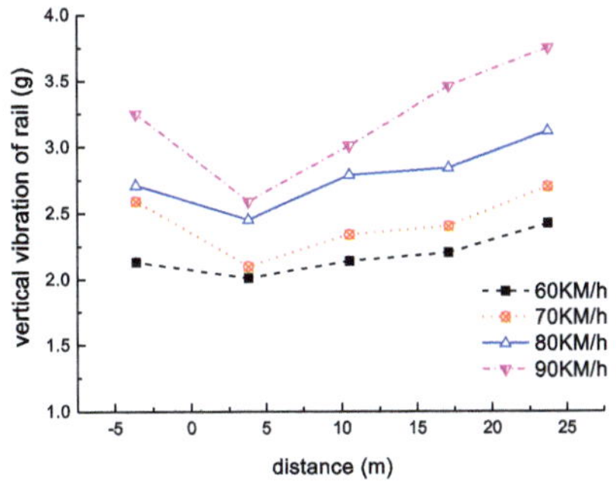

Fig. 15. Vibration characteristics of sleeper

6 Conclusion

To solve the track diseases in the transition zones of ballasted and ballastless track, a test section that the ballast bed with polyurethane polymer is established. In the test section, three bonded forms of ballast bed were adopted—full bonding section, partial bonding section and local bonding section. In order to detect the transition effect of ballast stiffness at the test section, the static test and dynamic test were carried out in the field. The results were as follows:

(1) Performance of polyurethane polymer in the transition zones can achieve the transition of ballast bed from low stiffness to high stiffness.
(2) The vertical stiffness of ballast bed with polyurethane polymer was increased more than 5 times, 3 times, 2 times. The vertical and horizontal resistance of ballast bed was increased more than 8 times, 4 times, 3 times.
(3) With the increase of track stiffness, the vertical force of wheel-rail and the vertical vibration of rail increase, the vertical displacement of rail reduce, the vertical displacement and vertical vibration of sleeper reduce.
(4) The vertical vibration of sleeper gradually increased with the increase of vehicle speed.

References

1. Sañudo, R., Dell'Olio, L., Casado, J.A., et al.: Track transitions in railways: a review. Constr. Build. Mater. **112**, 141–157 (2016). doi:10.1016/j.conbuildmat.2016.02.084
2. Lei, X.: Influences of track transition on track vibration due to the abrupt change of track rigidity. China Railway Sci. **27**(5), 42–45 (2006). doi:1001-4632(2006)05-0042-04. Springer
3. Li, W., Bian, X.: Dynamic performance of pile-supported bridge-embankment transition zones under high-speed train moving loads. Procedia Eng. **143**, 1059–1067 (2016). doi:10.1016/j.proeng.2016.06.101. (ICTG 2016)

4. Liu, Y., Guotang, Z., Qi, W., et al.: Dynamic analysis of ballasted-ballastless track transition section on high speed railway bridge. J. Vibr. Shock **34**(9), 76–81 (2015). doi:10.13465/j.cnki.jvs.2015.09.014
5. Woodward, P.K., Kennedy, J., Medero, G.: Three-dimensional track reinforcement using polymer geocomposites. In: Proceedings of the 2009 American Railway Engineering and Maintenance of Way Association (AREMA), Chicago, IIlinios, USA (2009)
6. Woodward, P.K., Kennedy, J., Mederol, G.M., et al.: Maintaining absolute clearances in ballasted railway tracks using in situ three-dimensional polyurethane geocomposites. Proc. Inst. Mech. Eng. Part F J. Rail Rapid Transit **226**(3), 257–271 (2011). doi:10.1177/0954409711420521
7. Woodward, P.K., Kennedy, J., Medero, G.M., et al.: Application of in-situ polyurethane geocomposite beams to improve the passive shoulder resistance of railway track. Proc. Inst. Mech. Eng. Part F J. Rail Rapid Transit **226**(3), 294–304 (2012). doi:10.1177/0954409711423460
8. Kennedy, J., Woodward, P.K., Banimahd, M., et al.: Railway track performance study using a new testing facility. Proc. Inst. Civil Eng. Geotech. Eng. **165**, 303–319 (2012). doi:10.1680/geng.10.00075]
9. Kennedy, J., Woodward, P.K., Medero, G.: Reducing railway track settlement using polyurethane polymer reinforcement of the ballast. Int. J. Constr. Build Mater. **44**, 615–625 (2013). doi:10.1016/j.conbuildmat.2013.03.002
10. Science and Technology Management Department of China Railway Corporation. TJ-GW116–2013 Provisional technical conditions of Polyurethane Polymer. China Railway Press, Beijing (2014)

Road Performance Prediction Model for the Libyan Road Network Depending on Experts' Knowledge and Current Road Condition Using Bayes Linear Regression

Abdussalam Heba(✉) and Gabriel J. Assaf

Ecole de Technologie Supérieure, 1100 Notre-Dame W, Montreal, QC, Canada
abdussalam.heba.1@ens.etsmtl.ca,
gabriel.assaf@etsmtl.ca

Abstract. The accurate prediction of rates of road deterioration is important in Pavement Management Systems (PMS), to ensure efficient and forward looking management and for setting present and future budget requirements. Libyan roads face increasing damage from the lack of regular maintenance. This reinforces the need to develop a system to predict road deterioration in order to determine optimal pavement intervention strategies (OIS). In a PMS, pavement deterioration can be modeled deterministically or probabilistically. This paper proposes a Bayesian linear regression method to develop a performance model in the absence of historical data; instead, the model uses expert knowledge as a prior distribution. As such, Libyan Road experts who have worked for a long time with the Libyan Road and Transportation Agency have been interviewed to develop input data to feed the Bayesian Model. A posterior distribution was computed using a likelihood function depending on road condition inspections in accordance with a pre-established protocol. The results were the pavement deterioration prediction models based on expert knowledge and a few on-site inspections.

Keywords: Pavement management systems · Pavement performance · International roughness index (IRI) · Bayesian linear regression

1 Introduction

A road pavement deteriorates under the combined action of traffic loading and environment, thus reducing the quality of the ride (Madanat et al. 1997). Models should be able to quantify the contribution of variables such as strength of pavement materials, traffic, and environmental conditions that are relevant to pavement deterioration (Ortiz-Garcia et al. 2006). PMS are commonly used to select maintenance strategies that result in lower project life cycle costs (Premkumar and Vavrik Haas 1994).

Modeling the performance of pavements is an important activity in pavement management, and many highway agencies have developed a variety of pavement performance models for use in their pavement management activities (Lethanh et al. 2014). This paper presents a methodology to develop new models for the various

J. Pombo and G. Jing (eds.), *Recent Developments in Railway Track and Transportation Engineering*, Sustainable Civil Infrastructures, DOI 10.1007/978-3-319-61627-8_12

pavement families in the Libyan road network in order to predict the condition of a given area of pavement. The predicted future condition of the pavements is used in estimating its remaining service life to failure, which will consequently be used to help find the best ways to intervene in the maintenance and rehabilitation activities for a given area of the network (Li et al. 1997).

There are mainly two basic kinds of performance models: deterministic and probabilistic (Kobayashi et al. 2012). The deterministic models predict a single number for the life of a pavement or its level of distress or any other measure of its condition. In contrast, the probabilistic models predict a distribution of such events. There are many deterministic models some of them are Mechanistic, Empirical-Mechanistic, Polynomial Constrained Least Squares, and S-Shaped Curve models (Hong et al. 2013). In general, probabilistic models include Bayesian and Markov process models. Bayesian modeling assigns a prior probability distribution to pavement condition based on experience; it then mixes it with the experimentation and data collection to predict the future condition (Hong and Prozzi 2006). Furthermore, Markov models can be used when pavement data is a sequence of conditions in which the probability of each condition depends only on the state attained in the previous condition (Prozzi and Madanat 2003; Li 2005).

1.1 Bayesian Model

The principle of Bayesian statistics lies in combining prior probabilities and likelihood with experimental outcomes to determine a post-experimental or posterior probability as shown in Fig. 1 (Pandis 2015a). The posterior distribution expresses what is known about a set of parameters based on both the sample data and prior knowledge (Han et al. 2014). In frequents statistics, it is often necessary to rely on large sample approximations by assuming asymptomatic normality. In contrast, Bayesian inferences can be computed exactly, even in highly complex situations (Jongsawat and Premchaiswadi 2010). This paper gives an account of basic uses of Bayes theorem and of the role and construction of prior densities. This follows by inference, dealing with analogues of confidence intervals, tests, approaches to model criticism, and model uncertainty (Gongdon 2003). Using the probability density function, Bayes model can be expressed as follows:

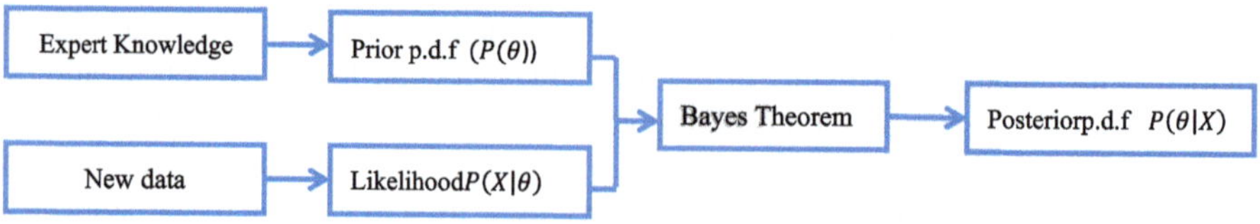

Fig. 1. Bayes General Concept

$$P(\theta|X) = \frac{P(X|\theta)P(\theta)}{P(X)} = \frac{P(X|\theta)P(\theta)}{\int P(X|\theta)P(\theta)d\theta} \tag{1}$$

A fundamental feature of the Bayesian approach to statistics is the use of prior information in addition to the sample data. A proper Bayesian analysis will always incorporate prior information, which will help to strengthen inferences about the true value of the parameter and ensure that any relevant information about it is not wasted (Nagaraja 2006).

1.1.1 Prior Knowledge $P(\theta)$

A fundamental feature of the Bayesian approach to statistics is the use of prior information in addition to the sample data. A proper Bayesian analysis will always incorporate prior information, which will help to strengthen inferences about the true value of the parameter and ensure that any relevant information about it is not wasted (Lunn et al. 2000).

1.1.2 Maximum Likelihood Estimation (MLE) $P(X|\theta)$

The maximum likelihood estimation (MLE) approach is one of the most important statistical methodologies for parameter estimation (Clark 2015). It is based on the fundamental assumption that the underlying probability distribution of the observations belongs to a family of distributions indexed by unknown parameters (Schwartz et al. 2013). The MLE estimator of the unknown parameters maximizes the likelihood function that corresponds to the probability distribution in the family that gives the observations the highest chance of occurrence. The MLE method starts from the joint probability distribution of the measured values $x_1, x_2, \ldots, x_n$. For independent measurements, this is given by the product of the individual densities $p(x|\theta)$, as in Eq. 2.

$$P(X|\theta) = p(x_1|\theta)p(x_2|\theta)\ldots p(x_n|\theta) = \prod\nolimits_{n}^{i=1} p(x_i|\theta) \tag{2}$$

1.1.3 Posterior Distribution $P(\theta|X)$

Posterior expresses what is known about a set of parameters based on both the sample data and prior information. Bayes theorem works as a mechanism for generating a posterior of any parameter and thereby mixes the prior knowledge with the likelihood. The first iteration production of the prior knowledge and the MLE will then be divided by $P(X)$, a normalizing factor, to normalize the distribution. When the posterior distribution $P(\theta|X)$ is in the same family as the prior probability distribution $P(\theta)$, the prior and posterior are then called conjugate distributions. Non-conjugate prior distributions can make interpretations of posterior inferences more difficult (Han et al. 2014).

1.2 The International Roughness Index (IRI)

The International Roughness Index (IRI) is an international standard for measuring road roughness longitudinally. The index measures pavement roughness in the wheel paths in terms of the number of rough meters per kilometer. The most common method uses a laser that is mounted on a specialized van. The laser is trained on the road surface, like a laser pointer. As the van drives along a road, the beam jumps unexpectedly at rough patches, just as a laser pointer; these jumps are measured and used for analysis (Mašović and Hajdin 2013). The lower the IRI number at a given speed, the smoother the ride felt by the road user. Moreover, this roughness statistic is suitable for any road surface type and covers all levels of roughness (Kobayashi et al. 2012). IRI can be treated as a random variable and therefore it can be described as a probability distribution (Shahin 2005). The main advantages of the IRI are that it is stable over time and transferable throughout the world. IRI can also be used as a measure of pavement conditions and the data can be easily shared between researchers. It can also be directly related to vehicle operating costs (Shahin 2005).

2 Methodology

The technique that will be used to estimate the road network deterioration is a variant of Bayesian Expert-Based probability matrices of deterioration. This technique can then be applied to road classes in Libya. This method depends on combining observed data with expert experience, using Bayesian linear regression techniques. The Bayesian prediction approach is the process of analyzing statistical models by using prior knowledge and observations as shown in Eq. 1 (Amador-Jimenez and Mrawira 2012). Bayesian linear regression adds more accuracy to the estimation of the parameters according to the International Roughness Index (IRI); this is because it covers the whole range of inferential solutions, rather than a point estimate and a confidence interval, as in classical regression (Davison 2008). The research methodology consists of three major steps. These are: interviewing experts in order to set up the prior distribution; inspecting road networks to estimate MLE; computing the posterior and predictive distributions for the IRI as can be seen in Fig. 2.

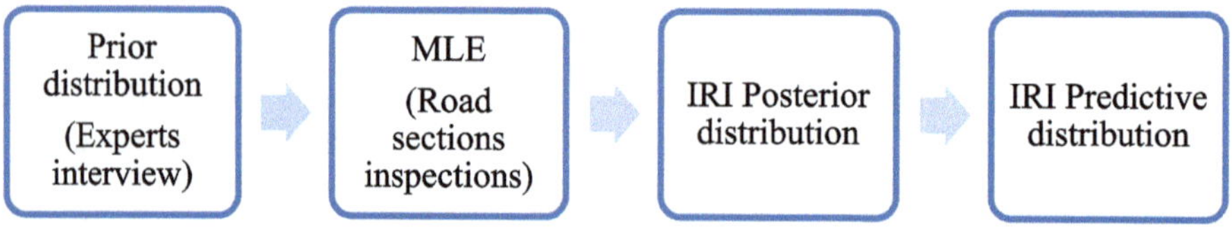

Fig. 2. Research methodology steps

2.1 Pavement Families Classification

Libya's most prominent natural features are the Mediterranean coast, the Sahara Desert and several highlands. As a result, the soil conditions were divided into three categories (low, medium, high). Moreover, the road network is exposed to two climate zones which are the hot-summer Mediterranean climate and the hot desert climate consequently; the network was categorized into north and south zones. Therefore, in this research, 3 loading levels, 3 soil conditions and 2 climate zones interact with each other

Table 1. Road network will be divided into Zones (North and south) interacting with traffic loads and soil conditions during a sequence of years.

		North Zone			South Zone		
		Load Level			Load Level		
		Low	Medium	High	Low	Medium	High
Soil Strength	Low	*Dataset1*	*Dataset2*	*Dataset3*	*Dataset10*	*Dataset11*	*Dataset12*
	Medium	*Dataset4*	*Dataset5*	*Dataset6*	*Dataset13*	*Dataset14*	*Dataset15*
	High	*Dataset7*	*Dataset8*	*Dataset9*	*Dataset16*	*Dataset17*	*Dataset18*

and result in 18 pavement families. These pavement families are then used to develop Bayesian linear regression prediction models for each family, as shown in Table 1.

Since loading and soil conditions are the most important factors that damage most pavement sections, they are often used as independent variables in developing equations that predict conditions. In many cases, they are combined with age as an independent variable. In most circumstances, agencies want to know in how many years a pavement will need intervention (Lethanh et al. 2014). Therefore, in some models, loads and types of soil are used as factors that affect the rate of deterioration of a road surfacc; in these cases they are both considered as independent variables. Road sections are selected using a random stratified sampling technique to avoid any biased estimations.

Soil strength is measured by a penetration test in accordance with the California Bearing Ratio test (CBR) which evaluates the subgrade strength of roads. Traffic loads are categorized as follows:

- Low: <50 vehicles/day
- Medium: 50–500 vehicles/day
- High: 500–2000 vehicles/day

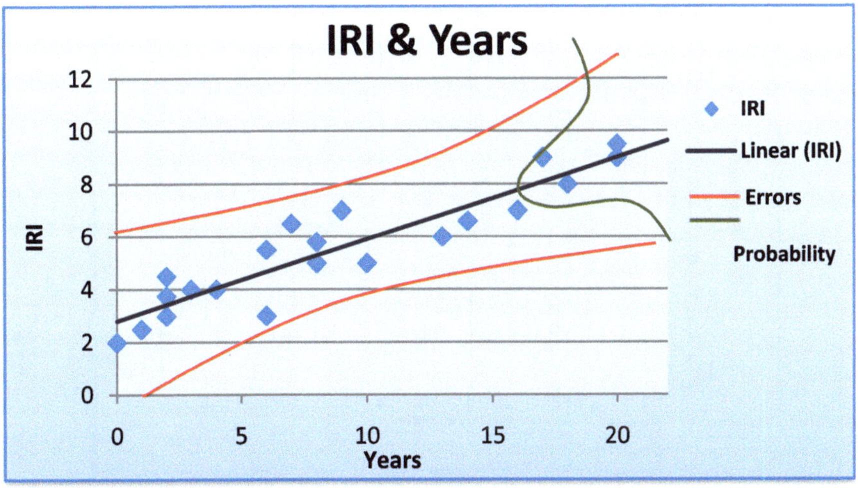

Fig. 3. The differences between linear regression and Bayesian linear regression

In general, these models are network-level deterioration models and not project-level deterioration models because the characteristics and the properties of the materials are not presently available in Libya.

2.2 Interviewing Experts (Prior) and Pavement Condition Inspections

The Bayesian statistical approach combines prior knowledge (experience) with field data. In highway engineering, new models are continually needed to better predict pavement performance or to run various PMS; however, it takes much time and expense to gather data about pavement performance. In such situations, the Bayesian approach is useful in short circuiting the data collection cycle. After gathering some data, which may not be sufficient to support meaningful classical regression, one can collect some expert judgment and combine the two sources of information into a relatively robust regression model. The expert judgment serves to bridge the gaps in field data.

It is obvious that a lot of valuable information can be obtained from people who have observed pavement performance throughout their careers. This professional and field staffs know what variables are contributing to pavement performance. They understand the functional relations of the variables. Their impressions on these relationships can be encoded and when combined with field data, these impressions can have profound impacts on the resulting posterior models. That is why; initial data has been collected by interviewing Libyan experts who have worked for many years on the development of the Libyan road network. Six engineers were interviewed using a standardized, open-ended interview technique; it is very structured and include a set protocol of questions and probes (Pandis 2015b).

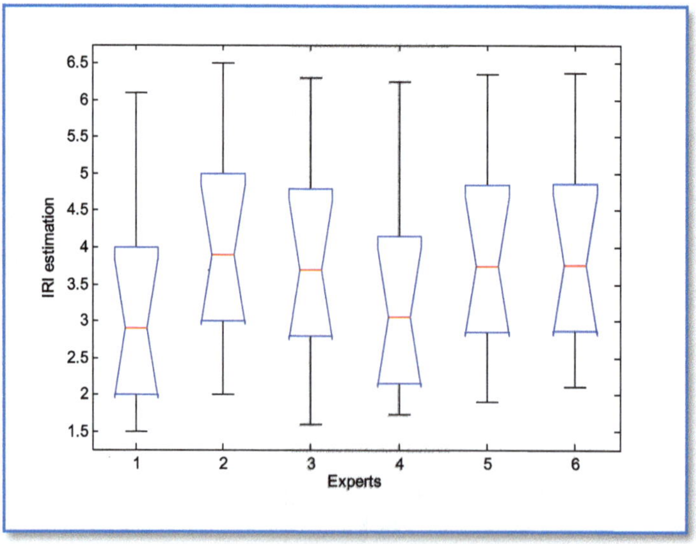

Fig. 4. ANOVA Experts' opinions comparison

Table 2. Two pavement families from north and south zones

	Pavement family 1		Pavement family 10	
	Prior	Pavement condition	Prior	Pavement condition
1	1.05	1.15	1.22	1.32
2	1.29	1.35	1.46	1.52
3	1.75	1.65	1.92	1.82
4	2.10	1.90	2.27	2.07
5	2.45	2.15	2.62	3.32
6	2.80	2.40	2.97	2.57
7	3.15	2.65	3.32	2.82
8	3.50	2.90	3.67	3.07
9	3.85	3.15	4.02	3.32
10	4.20	3.40	4.37	3.57
11	4.90	3.90	5.07	4.07
12	5.25	4.15	5.42	4.32
13	5.60	4.40	5.77	4.57
14	5.95	4.65	6.12	4.82
15	6.3	4.90	6.47	5.26
16	6.65	5.15	6.86	5.65
17	7.00	5.40	7.17	5.78
18	7.35	5.65	7.52	6.12
19	7.70	5.9	7.87	6.47
20	8.05	6.45	8.22	6.82
21	8.40	7.33	8.57	7.27

An analysis of variance was done before combining the experts' knowledge; this ensures that all experts' priors are consistent. From Fig. 4 and Table 3, there is no significant evidence to show that there is a difference in group means. As a result, the experts' opinions were considerably compatible; this means that all of the experts' knowledge about the roughness progression was close to each other. After that, an inspection of representative road sections from each the 18 families was conducted in Libya. The road deterioration was measured by the IRI on a subjective basis. Table 1 provides a summary of the characteristic of 18 families and codification of 18 associated databases. A sample of experts' interviews represents pavement families 1 and 10 is shown in Table 2.

Table 3. ANOVA results to investigate the differences between the means of experts' opinions

Source of variation	SS	df	MS	F	P-Value	F crit
Between groups	5.43	5	1.09	0.54	0.74	2.37
Within groups	120.27	60	2.00			
Total	125.70	65				

2.3 IRI Estimations

Roads deteriorate and their IRI drop gradually over time. This relationship can be represented using linear regression but, practically, road sections having the same zone, age, load, and soil strength conditions could still have a different rate of deterioration. Therefore, Bayesian linear regression is the appropriate technique wherein basic Bayesian philosophy is applied. This is because the Bayesian regression is a probabilistic approach that accounts for variability (refer to Fig. 3). As such, in Bayesian inference, MLE is considered to be point estimation. However, in Bayesian linear regression, productive probability around each inference of the IRI is probabilistically investigated (Amador-Jimenez and Mrawira 2012).

The research data required for IRI estimations has been divided into two main categories. The first category was extracted from the interviews with the experts. The second category is the MLE data; this data has been collected using the IRI as part of the road section inspections process, and has been done to measure the road deterioration. The MLE data is extracted and summarized as pairs of (t_i, IRI_j) where IRI represents the road section condition and t indicates the time.

$$Data = ((t_1, IRI_1), \ldots\ldots, (t_n, IRI_n)), 0 \le t_i \le 20, 0 \le IRI_j \le 14 \tag{3}$$

Therefore, the IRI_j is a model to be conditionally independent given the w vector which will be the prior distribution.

$$IRI_j \sim N\left(w^T t_i, a^{-1}\right), a > 0 \tag{4}$$

$$w \sim N\left(0, b^{-1} I\right), b > 0, w = (w_1, \ldots\ldots, w_d) \tag{5}$$

Where $a = \frac{1}{\sigma^2}$ is the precision factor, b *is* the covariance matrix; a and b are known, and w is a parameter vector with a Gaussian multivariate density.

2.3.1 The Posterior Distribution

The next step is to compute the posterior distribution on w given data. The t_i will be written as $\varphi(t_i) = (\varphi_1(t_1), \ldots., \varphi_n(t_1))$ in order to be able to model the nonlinearities of t_i. To compute the posterior, we need to calculate the MLE and then the predictive distribution.

2.3.2 Maximum Likelihood Estimation (MLE)

Given data:

$$D = (IRI_1, \ldots., IRI_n), IRI_i \in (0, 14) \tag{6}$$

D represents a sample from the *IRI* statistical population that has been collected from road section inspections. Then, the MLE is computed using the following formula:

$$P(D|w) \propto \exp\left(-\frac{a}{2}(IRI - Aw)^T(IRI - Aw)\right) \tag{7}$$

Where A is the design matrix and IRI is a value that we are going to predict, in a column vector form.

$$A = \begin{pmatrix} - & t_1^T & - \\ \vdots & \vdots & \vdots \\ - & t_n^T & - \end{pmatrix}, IRI = (IRI_1, \ldots\ldots, IRI_n)^T \tag{8}$$

2.3.3 Posterior

From the classical Bayesian definition, the posterior is proportional with the prior

$$P(w|D) \propto P(D|w)P(w) \tag{9}$$

After that, we replace the MLE expression in the posterior; this is shown as:

$$P(w|D) \propto \exp\left(-\frac{a}{2}(IRI - Aw)^T(IRI - Aw) - \frac{b}{2}w^T w\right) \tag{10}$$

With a little calculus we can express w in the form of a Gaussian distribution and call it a precision matrix:

$$P(w|D) = N(w|\mu, A^{-1})\ Where\ \mu = a\Lambda^{-1}A^T iri\,;\ \Lambda = aA^T A + bI \tag{11}$$

That shows us the Maximum Posterior (MAP) and MLE estimations of w, which are:

$$w_{MAP} = \left(A^T A + \frac{b}{a}I\right)^{-1} A^T iri \tag{12}$$

$$w_{MLE} = \left(A^T A\right)^{-1} A^T iri \tag{13}$$

2.3.4 Predictive Distribution

The predictive distribution is the conditional distribution of unobserved observations (prediction) given the collected data (Hong and Prozzi 2006). Our unobserved observation is the expert interview data; and the collected data is the data collected from road condition inspections, which can be expressed, in the following format:

$$P(iri|t, D) = \int P(iri|t, w)(w|t, D)dw \tag{14}$$

$$= \int N\left(iri|w^T t, a^{-1}\right) N\left(w|\mu, A^{-1}\right) dw \tag{15}$$

$$\propto \int exp(-\frac{a}{2}(iri - w^T t)^2 exp(-\frac{1}{2}(w - \mu)^T \Lambda(w - \mu)dw \quad (16)$$

This formula is then factored and put in a quadratic form as a function of w in a formula similar to the following general expression: $\int N(w|\ldots\ldots)g(iri)dw$ and then, since $g(iri)$ is not dependent on w, it comes out of the integral and $\int N(w|\ldots\ldots)dw$ integrates to 1.

After several algebraic steps, finalization of the predictive distribution is:

$$P(iri|t, D) = N\left(iri|u, \frac{1}{\lambda}\right) where u = \mu^T t \quad and \quad \frac{1}{\lambda} = \frac{1}{a} + t^T \Lambda^{-1} t \quad (17)$$

Finally, using mathematical expectation and Eq. (16) in all road section families, IRI will be estimated depending on:

- *iri* is the expert interview data,
- *t* is the time corresponding with road conditions,
- *D* is the data collected from road inspections.

3 Case Study (First Pavement Family)

To illustrate the effectiveness of this model, two data sources will be organized in a database file in order to be easily imported when the model is run. The first data source is the data extracted from interviewing Libyan experts. The interview data was used to formulate the prior probability distribution and some results of this interview are shown in Table 2. The second data source is from the selected road section inspection data (IRI). Table 1 shows two pavement families which have been chosen from each geographic zone (north & south). In this study, the model is applied only to the first pavement family, shown in Table 1.

3.1 Bayesian Regression Analysis

This section consists of all required steps to apply the Bayes regression on the collected data. The model has 1000 iterations using a combined prior to present all expert knowledge encoded in one model for each pavement family. A combined prior was selected to develop a single model for each pavement family. If each expert prior were analyzed separately, six separate posterior models would have to have been developed for each pavement family.

An analysis of variance was done before combining the experts' knowledge; this ensures that all experts' priors are consistent. From Fig. 4 and Table 3, there is no significant evidence to show that there is a difference in group means. As a result, the experts' opinions were considerably compatible; this means that all of the experts' knowledge about the roughness progression was close to each other.

WinBUGS was chosen as a programming platform; this is a free software available from the Biostatistics Unit of the Medical Research Council in the UK (Medical Research Council 2016). The WinBUGS program consists of three parts, all of which can be placed into a single file or as three separate files. The first part is the main program that is a string of computer code that lets WinBUGS know what the prior and likelihood of the model is. The second part is the data set that can be entered using matrixes in the same program or can be called from a file. The last part is the initial values that are used to start the algorithm. To estimate the parameters in Bayesian analysis, the prior distribution is multiplied by the likelihood; samples are then taken from the posterior distributions via an iterative Markov Chain Monte Carlo (MCMC) algorithm (Davison 2008).

3.2 Model Results (First Pavement Family)

The model combines data taken from the road condition inspections in accordance with a pre-established protocol and prior knowledge of the six experts who participated in the interviews. Once the model, the data, and the initial values have been specified, the program will be ready to be compiled and to run the MCMC algorithm. WinBUGS offers a Sample Monitor Tool panel which consists of a number of task icons as shown in Fig. 5. One of these tools is the stats tool; this gives a zoomed-out view of the entire posterior summary for the Bayes linear regression parameters, as illustrates in Table 4.

As a result of the Bayesian analysis, IRI predictive posterior models for the first pavement family were developed. The models have one independent variable and the predictive posterior equation is, as shown in Eq. (18), where α, β are the estimated parameters and t represents time in years.

$$IRI = \alpha + \beta t_i \tag{18}$$

Figure 6 shows the MCMC behaviour, where the chain appears to be moving around readily. This behaviour is called the dynamic trace because it will be continuously refreshed in real time if the model is updated. The probability distribution

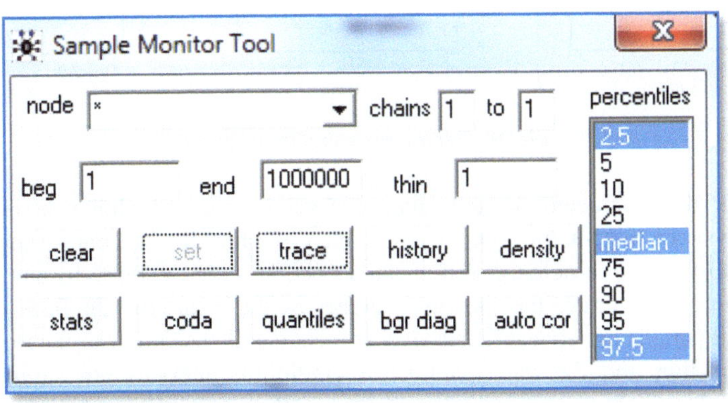

Fig. 5. Sample Monitor Tool

Table 4. Posterior summary for the Bayes linear regression parameters

Node	mean	sd	MC error	2.5%	median	97.5%	start	sample
alpha	0.7039	0.1896	0.01671	0.4934	0.6965	0.885	1	1000
beta	0.2850	0.01372	0.001211	0.2720	0.2857	0.3014	1	1000
tau	29.71	9.722	0.3757	12.81	28.91	51.02	1	1000

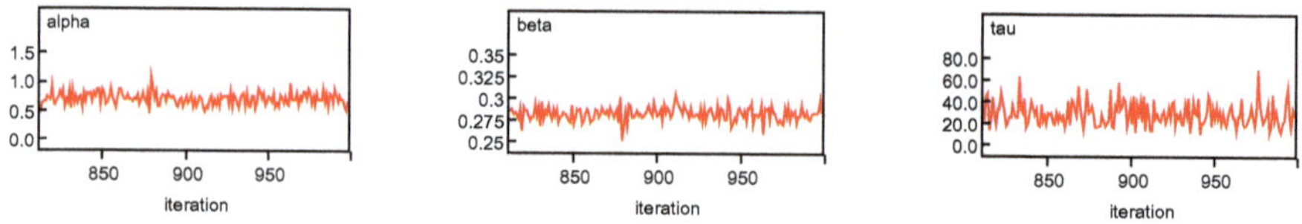

Fig. 6. Dynamic trace for the parameter outputs

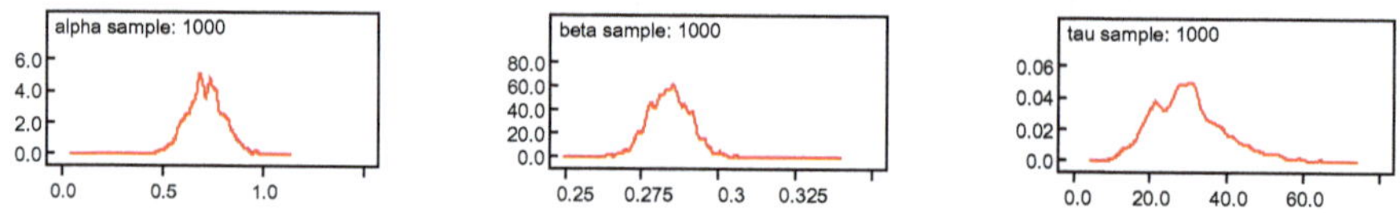

Fig. 7. Posterior densities for the model parameters

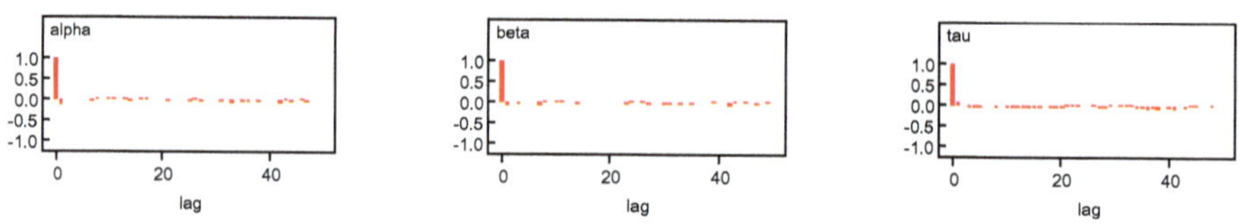

Fig. 8. Parameters autocorrelation functions

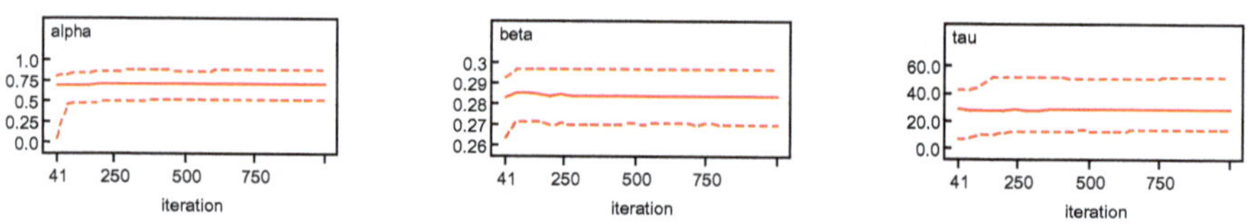

Fig. 9. Parameters running quantiles

densities of the parameters are displayed and summarized, as shown in Fig. 7. The basic analysis of the MCMC output is obtained by checking the convergence of the chain (history) and the autocorrelation (auto cor), as show in Figs. 10 and 8 respectively. Figure 9 illustrates the moving averages of the mean and the 95% credibility interval; all parameters appear stable over the course of the run. The parameters posterior estimation and the 95% parameters credibility intervals are summarized in Table 5. Table 6 shows a comparison between the actual IRI values and the IRI estimations which have been predicted by the model.

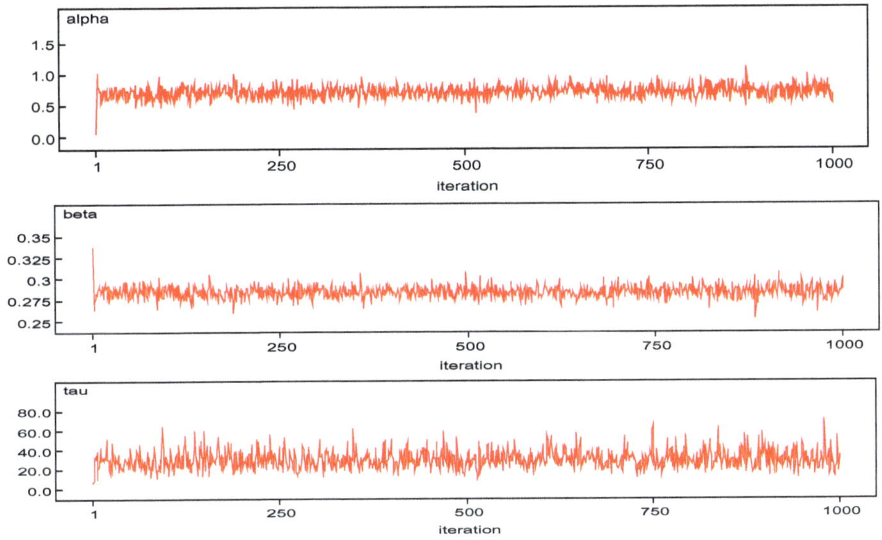

Fig. 10. Parameters trace history output

Table 5. Model result summary

Parameter	95% Credibility interval	Posterior mean	MLE
α	(0.4934, 0.885)	0.7039	0.6965
β	(0.2720, 0.3014)	0.2850	0.2857

Table 6. A comparison between actual IRI values and the model outputs

Road section	IRI	Predicted IRI
1	1.15	0.99
2	1.35	1.28
3	1.65	1.57
4	1.90	1.86
5	2.15	2.15
6	2.40	2.44

4 Conclusions

This paper has demonstrated how Bayesian linear regression modeling provides a more reliable framework for anticipation when historical data is not available. The linear regression model is undertaken within the boundaries of the Bayes inference approach, in order to investigate the parameters estimation errors probabilistically.

The paper consists of three major steps: interviews with experts to establish the prior distribution of the model; measuring the current road roughness using the IRI on a selected road sample; producing the posterior distribution followed by the predictive distribution. The result is a Bayesian linear regression with two parameters (α, β) where the model is expressed as $IRI = \alpha + \beta t_i$. Moreover, credibility intervals were accompanied with the parameter estimations; this increased the reliability of the domain estimation for the posterior probability distribution, as shown in Table 5.

This technique is highly recommended when developing a model to estimate pavement performance in the absence of historical data. Moreover, this method is not exclusive to the Libyan road network, but is applicable in any road network when the circumstances are similar especially in developing countries. The main disadvantage of this method is that, because it was developed for cases where there was no historical data, it does not have a mechanism to incorporate it. This required the researchers to adopt the approach that used interviews with experts instead; this was difficult and required a lot of consideration as to the type of questions necessary to extract the needed data for this research. Additionally, because of the lack of archived data, the authors have divided the road network into 18 families based on geographical location, traffic load and soil strength; this means that 18 models were developed, which was sometime unwieldy and oftentimes a lot of extra work.

Acknowledgments. I would like to thank the Libyan Road Agency as well as all the participating experts who have provided insight and expertise, without which this project would not have been possible.

References

Amador-Jimenez, L.E., Mrawira, D.: Bayesian regression in pavement deterioration modeling: revisiting the AASHO road test rut depth model. Infraestruct. Vial **25**, 28–35 (2012)

Clark, M.: Bayesian Basics, *a conceptual introduction with application in R and Stan*. 2nd 13 Jun 2015. https://m-clark.github.io/docs/IntroBayes.html

Davison, A.C.: Statistical Models. Cambridge University Press, New York (2008)

Gongdon, P.: Applied Bayesian Modelling. Wiley, West Sussex (2003)

Haas, R.: Pavement Management Guide. Transportation Association of Canada, Ottawa (1977)

Haas, R.: Modern Pavement Management. Krieger Pub Co., Malabar Florida (1994)

Han, D., et al.: Application of Bayesian estimation method with Markov hazard model to improve deterioration forecasts for infrastructure asset management. KSCE J. Civil Eng. **18**(7), 2107–2119 (2014)

Hong, F., Prozzi, J.A.: Estimation of pavement performance deterioration using Bayesian approach. J. Infrastruct. Syst. **12**(2), 77–86 (2006)

Hong, T., et al.: Infrastructure asset management system for bridge projects in South Korea. KSCE J. Civil Eng. **17**(7), 1551–1561 (2013)

Jongsawat, N., Premchaiswadi, W.: Bayesian network inference with qualitative expert knowledge for decision support systems. pp. 3–8 (2010)

Kobayashi, K., et al.: A statistical deterioration forecasting method using hidden Markov model for infrastructure management. Transp. Res. Part B: Methodol. **46**(4), 544–561 (2012)

Lethanh, N., et al.: Optimal intervention strategies for multiple objects affected by manifest and latent deterioration processes. Struct. Infrastruct. Eng. **11**(3), 389–401 (2014)

Li, N., et al.: Development of a new asphalt pavement performance prediction model. Can. J. Civil Eng. **24**(4), 547–559 (1997)

Li, Z.: A Probabilistic and Adaptive Approach to Modeling Performance of Pavement Infrastructure. Faculty of the Graduate School, University of Texas at Austin. **Ph.d**. (2005)

Lunn, D., et al.: WinBUGS - A Bayesian modelling framework: Concepts, structure, and extensibility. Stat. Comput. **10**(4), 325–337 (2000)

Madanat, S.M., et al.: Probabilistic infrastructure deterioration models with panel data. J. Infrastruct. Syst. **3**(1), 4–9 (1997)

Mašović, S., Hajdin, R.: Modelling of bridge elements deterioration for Serbian bridge inventory. Struct. Infrastruct. Eng. **10**(8), 976–987 (2013)

Medical Research Council: WinBugs Installation (2016). Retrieved 30 Sept 2016, http://www.mrc-bsu.cam.ac.uk/software/bugs/

Nagaraja, H.N.: Inference in hidden markov models. Technometrics **48**(4), 574–575 (2006)

Ortiz-Garcia, J.J., et al.: Derivation of transition probability matrices for pavement deterioration modeling. J. Transp. Eng. **132**(2), 141–161 (2006)

Pandis, N.: The sampling distribution. Am. J. Orthod. Dentofac. Orthop. **147**(4), 517–519 (2015a)

Pandis, N.: Statistical inference with confidence intervals. Am. J. Orthod. Dentofac. Orthop. **147**(5), 632–634 (2015b)

Premkumar, L., Vavrik, W.R.: Enhancing pavement performance prediction models for the Illinois Tollway System

Prozzi, J., Madanat, S.: Incremental nonlinear model for predicting pavement serviceability. J. Transp. Eng. **129**(6), 635–641 (2003)

Schwartz, J., et al.: Improved maximum-likelihood estimation of the shape parameter in the Nakagami distribution. J. Stat. Comput. Simul. **83**(3), 434–445 (2013)

Shahin, M.Y.: Pavement Management for Airports, Roads and Parking Lots. Springer Science, New York (2005)

Author Index

J. Pombo and G. Jing (eds.), *Recent Developments in Railway Track and Transportation Engineering*, Sustainable Civil Infrastructures, DOI 10.1007/978-3-319-61627-8

MIX
Papier aus verantwortungsvollen Quellen
Paper from responsible sources
FSC® C105338

If you have any concerns about our products,
you can contact us on
ProductSafety@springernature.com

In case Publisher is established outside the EU,
the EU authorized representative is:
Springer Nature Customer Service Center GmbH
Europaplatz 3, 69115 Heidelberg, Germany

Printed by Libri Plureos GmbH
in Hamburg, Germany

MIX
Papier aus verantwortungsvollen Quellen
Paper from responsible sources
FSC® C105338

If you have any concerns about our products,
you can contact us on
ProductSafety@springernature.com

In case Publisher is established outside the EU,
the EU authorized representative is:
Springer Nature Customer Service Center GmbH
Europaplatz 3, 69115 Heidelberg, Germany

Printed by Libri Plureos GmbH
in Hamburg, Germany